SpringerBriefs in Electrical and Computer Engineering

Control, Automation and Robotics

Series Editors

Tamer Başar
Antonio Bicchi
Miroslav Krstic

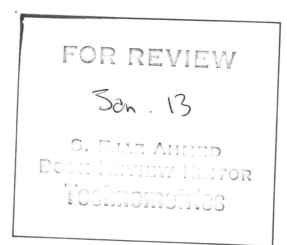

For further volumes:
http://www.springer.com/series/10198

Mathukumalli Vidyasagar

Computational Cancer Biology

An Interaction Network Approach

 Springer

Mathukumalli Vidyasagar
Bioengineering Department
The University of Texas at Dallas
Richardson
USA

ISSN 2192-6786   ISSN 2192-6794   (electronic)
ISBN 978-1-4471-4750-3   ISBN 978-1-4471-4751-0   (eBook)
DOI 10.1007/978-1-4471-4751-0
Springer London Heidelberg New York Dordrecht

Library of Congress Control Number: 2012950853

Printed on acid-free paper

Springer is part of Springer Science+Business Media (www.springer.com)

To Mike White

Preface

During the first four decades after completing my Ph.D., I was a 'card-carrying control theorist'. Sometime toward the end of this period I became aware that there was a major revolution taking place in biology, whereby it became pretty easy for anyone with a sufficiently large budget to generate massive amounts of raw data. The challenge, I was repeatedly told, was to process these data to identify patterns and draw valid conclusions. Accordingly I tried to read several books and articles on bioinformatics and computational biology, but never felt sufficiently enlightened as to what the underlying issues were, or enthused to pursue the subject. After I moved to the University of Texas at Dallas, I met Professor Michael A. White, Scientific Director of the Harold Simmons Cancer Center at the famed UT Southwestern Medical Center, also in Dallas. I can say this was a transformative event in my life. Mike is a rarity among biologists I have met, someone who is able to convey a broad picture of biological issues, and also to have an open mind toward those from alien cultures such as myself. Through intense interactions with Mike over the past two plus years, I have been able to formulate several statistical and algorithmic problems that are both interesting and challenging to those with a systems and control background, as well as useful to cancer biologists. While I have a long way to go, at least I have commenced on what promises to be a richly rewarding journey.

Another stroke of luck was Edwin Chong, in his capacity as General Chair of the 2011 joint Conference on Decision and Control and the European Control Conference, inviting me to give a plenary lecture. Under normal conditions, the plenary lecturer just shows up and gives his talk, hands over a copy of the slides, and that's that. But since this was a joint CDC-ECC, I had to prepare a proper journal article. This forced me to channel my meandering thoughts into something more coherent, and resulted in the tutorial paper [1]. So when Springer-Verlag asked me to write a brief monograph, it was natural to expand that paper into the present work, by elaborating on some of the ideas and including subsequent research. Due to the page limitations, even this work describes only a part of the research that my students and I are currently carrying out.

During the two years plus that Mike and I have been talking, I have picked up enough of the rudiments of cancer biology that I can now, with some confidence, act as an ambassador of the cancer biology community to the control theory community. I have not yet had the courage to try it in the other direction. Nonetheless, for his definitive role in converting me into at least a passable imitation of a computational cancer biologist, I take great pleasure in dedicating this book to Mike.

I acknowledge with gratitude financial support from the National Science Foundation Award #1001643, the Cecil & Ida Green Endowment at UT Dallas, and from the Harold Simmons Comprehensive Cancer Center at UT Southwestern Medical Center.

Hyderabad and Dallas, September 2012   Mathukumalli Vidyasagar

Reference

1. Vidyasagar, M.: Probabilistic methods in cancer biology. Eur. J. Control **17**(5–6), 483–511 (2011)

Contents

Editors' Bios

Tamer Başar is with the University of Illinois at Urbana-Champaign, where he holds the academic positions of Swanlund Endowed Chair, Center for Advanced Study Professor of Electrical and Computer Engineering, Research Professor at the Coordinated Science Laboratory, and Research Professor at the Information Trust Institute. He received the B.S.E.E. degree from Robert College, Istanbul, and the M.S., M.Phil, and Ph.D. degrees from Yale University. He has published extensively in systems, control, communications, and dynamic games, and has current research interests that address fundamental issues in these areas along with applications such as formation in adversarial environments, network security, resilience in cyber-physical systems, and pricing in networks.

In addition to his editorial involvement with these *Briefs*, Başar is also the Editor-in-Chief of *Automatica*, Editor of two Birkhäuser Series on *Systems & Control* and *Static & Dynamic Game Theory*, the Managing Editor of the *Annals of the International Society of Dynamic Games* (ISDG), and member of editorial and advisory boards of several international journals in control, wireless networks, and applied mathematics. He has received several awards and recognitions over the years, among which are the Medal of Science of Turkey (1993); Bode Lecture Prize (2004) of IEEE CSS; Quazza Medal (2005) of IFAC; Bellman Control Heritage Award (2006) of AACC; and Isaacs Award (2010) of ISDG. He is a member of the US National Academy of Engineering, Fellow of IEEE and IFAC, Council Member of IFAC (2011–2014), a past president of CSS, the founding president of ISDG, and president of AACC (2010–2011).

Antonio Bicchi is Professor of Automatic Control and Robotics at the University of Pisa. He graduated at the University of Bologna in 1988, and was a postdoc scholar at M.I.T. A.I. Lab between 1988 and 1990.

His main research interests are in:

- dynamics, kinematics, and control of complex mechanichal systems, including robots, autonomous vehicles, and automotive systems;

- haptics and dextrous manipulation; and
- theory and control of nonlinear systems, in particular hybrid (logic/dynamic, symbol/signal) systems.

He has published more than 300 papers on international journals, books, and refereed conferences.

Professor Bicchi currently serves as the Director of the Interdepartmental Research Center "E. Piaggio" of the University of Pisa, and President of the Italian Association or Researchers in Automatic Control. He has served as Editor-in-Chief of the Conference Editorial Board for the IEEE Robotics and Automation Society (RAS), and as Vice President of IEEE RAS, Distinguished Lecturer, and Editor for several scientific journals including the *International Journal of Robotics Research*, the *IEEE Transactions on Robotics and Automation*, and *IEEE RAS Magazine*. He has organized and co-chaired the first World Haptics Conference (2005), and Hybrid Systems: Computation and Control (2007). He is the recipient of several best paper awards at various conferences, and of an Advanced Grant from the European Research Council. Antonio Bicchi has been an IEEE Fellow since 2005.

Miroslav Krstic holds the Daniel L. Alspach chair and is the founding director of the Cymer Center for Control Systems and Dynamics at University of California, San Diego. He is a recipient of the PECASE, NSF Career, and ONR Young Investigator Awards, as well as the Axelby and Schuck Paper Prizes. Professor Krstic was the first recipient of the UCSD Research Award in the area of engineering and has held the Russell Severance Springer Distinguished Visiting Professorship at UC Berkeley and the Harold W. Sorenson Distinguished Professorship at UCSD. He is a Fellow of IEEE and IFAC. Professor Krstic serves as Senior Editor for *Automatica* and *IEEE Transactions on Automatic Control* and as Editor for the Springer series *Communications and Control Engineering*. He has served as Vice President for Technical Activities of the IEEE Control Systems Society. Krstic has co-authored eight books on adaptive, nonlinear, and stochastic control, extremum seeking, control of PDE systems including turbulent flows, and control of delay systems.

Chapter 1
The Role of System Theory in Biology

Abstract In this chapter we introduce the reader to current methods for generating biological data, including such topics as micro-array (or gene expression) studies, ChIP-seq studies, siRNAs, and micro-RNAs. Special features of biological data that necessitate the development of new algorithms are highlighted, such as the lack of standardization in experimental procedures that lead in turn to broad variability of the data sets.

Keywords Micro-array · ChIP-seq · siRNA · Micro-RNA

1.1 Introduction

Recent advances in experimental techniques, coupled with a dramatic reduction in the cost of experimentation, now permit the biology community to generate vast amounts of raw data at an affordable cost. However, this is only the beginning. It is necessary to analyze the data, so as to convert raw data into information, and information into actionable knowledge. This conversion would benefit enormously by the use of techniques from probability theory and statistics, graph theory, and machine learning, to mention just a few pertinent areas. However, it is not always possible to make meaningful contributions to biology by applying 'off the shelf' techniques from these areas. Biological data sets have some unique features that require the development of *ab initio* solution methodologies. For instance, in many machine learning problems that occur in engineering, such as recognizing handwritten characters or faces from images, the number of samples is a few orders of magnitude larger than the number of features. In biological problems however the situation is exactly the inverse: the number of features is one or two orders of magnitude more than the number of samples. Because of this fact, traditional theorems in machine learning that tell us what happens as the number of training samples approaches infinity are simply irrelevant in a biological context. Similarly, when a large number

M. Vidyasagar, *Computational Cancer Biology*, SpringerBriefs in Control,
Automation and Robotics, DOI: 10.1007/978-1-4471-4751-0_1, © The Author(s) 2012

of samples are available of a small number of random variables, it would be possible to deduce their joint distribution to a high level of accuracy and confidence. However, when there are relatively few samples of a large number of random variables, it is not possible to deduce their joint distribution. Instead we must be content with identifying families of joint distributions that are *consistent* with the data, but are by no means uniquely determined by the data. Missing measurements are far more prevalent in biological data than they are in engineering. Finally, in many ways the biological community is charging ahead with generating huge amounts of data even before the underlying measurement technologies have been standardized. As a result, when different vendors provide 'probes' that ostensibly measure the same quantity, often there is no resemblance whatsoever between the measurements generated by the various platforms. Even if one were to use the same vendor's apparatus consistently, the measured values are still subject to 'batch effects', either in the form of drift or additive and/or multiplicative noise. At present, there are massive public databases that serve as repositories of various data sets that have been generated by individual laboratories. However, the data that is deposited into these repositories is not normalized for platforms, batch effects, and the like. Thus, while it may appear that there is a great deal of data available, often it is not internally consistent and/or is of poor quality. These are but a small sample of the unique features of biological data sets. Thus any algorithms developed to handle biological data must incorporate provisions for coping with all of these phenomena.

The requirements of understanding these unique features, and appreciating the kinds of insights that biologists seek to derive from their data, provide a great opportunity to the engineering community to make a significant impact on biology. This is especially true of the systems and control community, because biological problems do not neatly fall into one clearly defined category of mathematics; rather, they cut across several categories, and it is incumbent on the problem solver to find the appropriate mathematical model(s) and then to tune the solution to the biologists' requirements. Since by nature control theory draws upon a wide variety of techniques from mathematics, it would be natural to apply the 'systems approach' to biological problems.

The interplay between mathematical modeling and experimental biology dates back several decades, and 'theoretical biology' has been a well-accepted discipline for a very long time, even if the name of the area keeps changing with time. In recent years, many persons whose primary training was in the systems and control area have moved into biology and have made many significant contributions, and continue to do so. This is illustrated by the fact that in recent years there have been two special issues within the controls community that are devoted to systems biology [1, 2]. It would be impossible to create a comprehensive description of all these contributions, and in any case, that is not the objective of this work. Rather, the objective of this brief monograph is to present a snapshot of one specific research problem in cancer biology to which methods from probability and statistics may be fruitfully applied. In that sense, the scope of the work is voluntarily limited.

Biology is a vast subject and cancer biology is a large part of this vast subject. Moreover, our understanding of this topic is constantly shifting, and there are very few 'settled' theories.[1] Hence the choice of the specific topic discussed here is dictated by the fact that it presents some interesting challenges in probability theory and statistics, and of course by the author's personal tastes. The hope is that this monograph would serve to present the flavor of this subject, and thus motivate interested readers to explore the literature further.

1.2 Some Facts and Figures About Cancer

Cancer is one of the oldest diseases known to man. A papyrus popularly known as the 'Ebers papyrus', dating to around 1500 BCE, recounts a 'tumor against the God Xenus' and suggests 'Do thou nothing there against'. A part of the Ebers papyrus is reproduced in Fig. 1.1.

The currently used cancer-related terms come from both Greek and Latin. In ancient times, the Greek word 'karkinos', meaning 'crab', was used to refer to the crab nebula as well as the associated zodiac sign. Supposedly Hippocrates in c.420 BCE used the word 'karkinos' to describe the *disease*, and 'karkinoma' to describe a cancerous *tumor*. One can surmise that he was influenced by the crab-like appearance of a cancerous tumor, with a hard and elevated central core and lines radiating from the core. Subsequently the name for the disease was changed to the Latin word 'cancer' which also meant 'crab', while the name for the tumor was transliterated into the Roman alphabet as 'carcinoma.' In recent times, the pronunciation of the second 'c' got 'mutated' to the 's' sound instead of the 'k' sound.

The web site [5] contains a wealth of statistics about the incidence, survival rates etc. of cancer, some of which are given below. Today cancer is the second leading cause of death worldwide, after heart failure, and accounts for roughly 13 % of all deaths. Contrary to what one may suppose, cancer occupies the second place even in developing countries. In the USA, about 1.5 million persons will be diagnosed with cancer in a year, while around 570,000 will die from it.

Over the years, quite substantial success has been realized in the treatment of *some forms* of cancer. This can be quantified by using the so-called five-year relative survival rate (RSR), which is defined as the ratio of the fraction of those with the disease condition that survive for five years, divided by the same number for the general population. To illustrate, suppose we start with a cohort of 1,000 persons. Assuming a mortality rate of 2 % per year for the general population, after five years roughly 900 of the original population will survive (rounding off the numbers for illustrative purposes). Now suppose that amongst a cohort of 1,000 persons with a particular form of cancer, only 360 survive for five years. Then the five-year RSR is $360/900 = 40\%$.

---

[1] For a very readable and yet scientifically accurate description of how theories about the onset and treatment of cancer have evolved over the past hundred years or so, see [3].

Fig. 1.1 The Ebers Papyrus [4]

Table 1.1 Relative survival rates over the years

Primary site	5-year RSR 1950–1954	5-year RSR 1999–2006
All sites	35	69.1
Childhood	20	82.9
Leukemia	10	56.2
Hodgkin lymphoma	30	87.7
Breast	60	91.2
Prostate	43	99.9
Pancreas	1	5.8
Liver	1	13.7
Lung	6	16.8

Table 1.1 shows the RSR for various forms of cancer, in 1954 and in 2006.

The consensus amongst cancer researchers is that, except for childhood cancer, the vast improvements in RSR in other forms of cancer are mostly an artefact of earlier detection and not an indication of improved treatment. Also, in some forms of cancer, such as pancreas, liver, and lung, the RSR figures have remained stubbornly stuck at very low levels. Not surprisingly, these diseases form a primary focus of cancer studies.

1.3 Advances in Data Generation

In this section we describe a few of the more widely used methods of generating experimental data at the molecular level. Molecular data complements clinical data, which is obtained by studying the outcomes of various therapies given to patients.

Finding reliable methodologies for predicting clinical outcomes from molecular data is one of the most widely sought-after goals of computational approaches to cancer biology.

1.3.1 Genome Sequencing

All living things propagate genetic information through DNA, which stands for Deoxyribonucleic acid, and is the fundamental building block of life. DNA is made up of four nucleic acids, which are usually denoted by the initial letter of the base that they contain, namely: A for Adenine, C for Cytosine, G for Guanine and T for Thymine. The genome of an organism is just an enumeration of its DNA. For present purposes, one can think of the genome as just an enormously long string over the four-symbol alphabet $\{A, C, G, T\}$. Thus the genome of an organism is its 'digital' description at the most basic level. Genes are the operative part of the DNA that produce proteins and thus sustain life. When the first 'complete' human genome, consisting of nearly 3.3 billion base pairs[2] was published in 2001 [6, 7], the project cost more than $3 billion and took nearly ten years; on top of that, it was only a 'draft' in that its error rate was roughly 2 %. Today there are commercial companies that promise to sequence a complete human genome at a cost of $5,000 or so per genome, or sell the equipment to do so. If one is not interested in the entire genome but only very specific loci where mutations are believed to be more common and/or impactful, the cost would be even lower. This is an impressive reduction of several orders of magnitude in both the cost and the time needed. Moreover, the cost and time continue to decrease at rates comparable to or faster than the fabled Moore's law of semiconductors. Because of these technological advances, it is now feasible to sequence literally tens of thousands of cancer tissues that are available at various research laboratories. The National Institutes of Health (NIH) has embarked upon a very ambitious project called TCGA (The Cancer Genome Atlas) whose ultimate aim is to obtain a comprehensive molecular characterization of every single cancerous tissue that is currently available to it, including the exome sequence, DNA copy number, promoter methylation, as well as expression analysis of messenger RNA and micro RNA [8].[3] By comparing (wherever possible) the DNA sequence of tumor tissue with the normal tissue of the same individual, it is possible to isolate many mutations. One of the main challenges in cancer is to distinguish between mutations that are causal from those that are coincidental.

---

[2] The phrase 'base pair' refers to the fact that DNA consists of two strands running in opposite directions, and that the two strands have 'reverse complementarity'—A occurs opposite T and C occurs opposite G.

[3] The reader is referred to any standard text on cell biology to gain an understanding of the various terms used here.

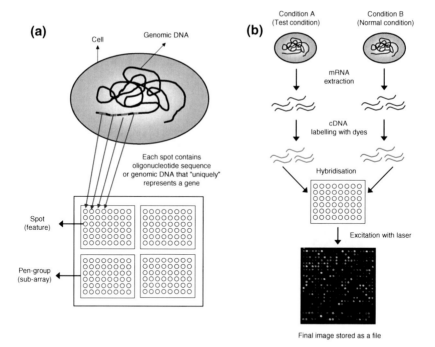

Fig. 1.2 Microarray experimentation [9, Fig. 1]

1.3.2 Microarray Experiments

In this subsection we briefly describe the basic principles behind a microarray experiment. Much of the material below, including Fig. 1.2, is taken from an excellent
overview [9]. Microarray experiments are a means of simultaneously measuring the
activity level of various genes under common set of specified experimental conditions. The microarray itself is a glass plate consisting of a few hundred cavities,
usually referred to as spots or wells. Each spot or well contains quantities of a different DNA fragment. Often the content of each well consists of replicates of a unique
gene, which is why such studies are also often referred to as gene expression studies.
By combining several arrays, it is possible in a single experiment to study the behavior of tens of thousands of genes. If every (or nearly every) gene in a cell is included
in the microarray study, the study is said to be 'genome-wide'. During the early
days of microarrays, individual laboratories prepared their own arrays, but now it is
almost universally the case that microarrays are either bought 'off the shelf' or, if the
spots are to be custom-made, ordered from commercial vendors. Each experiment
determines the difference between the number of RNA molecules corresponding to
a particular gene in two different kinds of cells, for example, normal tissue and

cancerous tissue; this comparison is carried out for a large number of genes. One of the experiments (say normal tissue) is referred to as the reference while the other (say cancerous tissue) is referred to as the test condition.

For the next step, we quote from [9]:

> First, RNA is extracted from the cells. Next, RNA molecules in the extract are reverse transcribed into cDNA by using an enzyme reverse transcriptase and nucleotides labelled with different fluorescent dyes. For example, cDNA from cells grown in condition A may be labelled with a red dye and from cells grown in condition B with a green dye. Once the samples have been differentially labelled, they are allowed to hybridize onto the same glass slide. At this point, any cDNA sequence in the sample will hybridize to specific spots on the glass slide containing its complementary sequence. The amount of cDNA bound to a spot will be directly proportional to the initial number of RNA molecules present for that gene in both samples. Following the hybridization step, the spots in the hybridized microarray are excited by a laser and scanned at suitable wavelengths to detect the red and green dyes. The amount of fluorescence emitted upon excitation corresponds to the amount of bound nucleic acid. For instance, if cDNA from condition A for a particular gene was in greater abundance than that from condition B, one would find the spot to be red. If it was the other way, the spot would be green. If the gene was expressed to the same extent in both conditions, one would find the spot to be yellow, and if the gene was not expressed in both conditions, the spot would be black. Thus, what is seen at the end of the experimental stage is an image of the microarray, in which each spot that corresponds to a gene has an associated fluorescence value representing the relative expression level of that gene.

In the above explanation, it is clear that the output of the microarray is a ratio between the number of RNA molecules present in the test and the reference conditions. As such, it is a nonnegative number. However, greater insight might be obtained by taking the logarithm and/or subtracting a bias value, in which case the normalized output can indeed be a negative number.

There are several problems associated with microarray studies as they are currently carried out. In most microarrays, each gene that is studied is associated with at least one 'probe', but sometimes more than one probe. The association between probes and gene expression is one of the many challenges in such studies. When the expression level of the same gene is measured on platforms from two different vendors, often there is no relationship whatsoever between the outputs of the two platforms. To elaborate, suppose one were to plot the outputs of the expression level of the same gene in the same tissue, as measured by two different probes from two platforms, against each other, across a patient population. Then it is not uncommon for the plot to resemble random noise. Figure 1.3 shows a plot of the expression levels of the gene C11orf76 as measured by two different probes from two different vendors, on a collection of ovarian cancer tissues surgically removed from patients. The raw data for this plot comes from the TCGA ovarian cancer study [10].[4] Each point represents a pair of values from the same tissue as measured by the two platforms. Evidently there is no coherence between the two measurements.

Even with two probes for one gene from the same vendor, the correlation between the outputs across a patient population is not always decisive. We will return to this topic again in Chap. 4 on future research directions.

---

[4] Plot generated by my student Burook Misganaw.

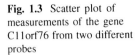

Fig. 1.3 Scatter plot of measurements of the gene C11orf76 from two different probes

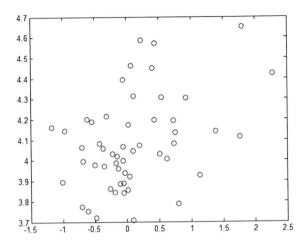

1.3.3 Chromatin Immunoprecipitation and ChIP-seq

Roughly speaking, chromatin immunoprecipitation (ChIP) is an experimental technique for predicting which parts of DNA in a cell might interact with a particular protein of interest. This technique is frequently used when the protein of interest is a transcription factor, that is, a gene that regulates the expression level of other genes. The 'pure' ChIP technique is rife with false positives, and routinely turns up thousands of possible target genes, most of which are false. In earlier days, the false positives were trimmed using microarray analysis, thus leading to a method called ChIP-chip. Nowadays a more accurate technique known as ChIP-seq is used. ChIP-seq looks for possible binding sites in the vicinity of each of the predicted target genes for the transcription factor. Obviously this requires very precise sequencing of the DNA (whence the suffix 'seq'), but the predicted target genes are more reliable than in ChIP-chip. This last step of eliminating false positives is referred to as 'peak calling' in the biology literature and is basically a signal processing problem with biological overtones. This too is perhaps an area where engineers can make a contribution. See [11] for a tutorial on chromatin immunoprecipitation, [12] for a tutorial on ChIP-seq, and [13] for one of several existing methods for peak calling. In fact in our work on reverse engineering a gene interaction network for lung cancer using gene expression data from cell lines (Sect. 3.6), the peak calling method used to identify potential downstream target genes is indeed that in [13].

1.3.4 Small Interfering RNAs

When cells reproduce, DNA gets converted to RNA (Ribonucleic acid) which in turn produces any of the roughly 100,000 proteins that sustain life. Unlike DNA which is a chemically stable molecule, RNA is somewhat unstable and can be thought of as an intermediary stage. The conversion of DNA to RNA (transcription) and of RNA to proteins (translation) is usually referred to as the 'central dogma' of biology; it was enunciated by Crick [14]. siRNA stands for 'small interfering RNA' (the expansion 'silencing RNA' is also used). siRNAs are small double-stranded RNA molecules, just 20–25 nucleotides long, that play a variety of roles in biology. For our purposes, the most important role is that each siRNA gets involved in the RNAi (RNA interference) pathway, and interferes with the expression of a specific gene. The first siRNAs that were discovered were naturally occurring; however, nowadays it is common to synthesize siRNAs in the laboratory, for targeted silencing of any gene of one's choice. Thus, for example, one can take a cancerous cell line that is kept alive in a laboratory ('immortalized'), apply a specific siRNA, and see whether or not the application of the siRNA causes the cell line to die out. If the answer is 'yes', then we conclude that the gene which is silenced by that specific siRNA plays a key role in the survival of the cancerous cell.

1.3.5 Micro-RNAs

Micro-RNAs, the native equivalent of siRNAs, are relatively short RNA molecules, roughly 20 nucleotides long, that bind to messenger RNA and inhibit some part of the translation aspect. At present there are about known 1,500 micro-RNAs. As a gross over-simplification, it can be said that each micro-RNA inhibits the functioning of more than one gene, while each gene is inhibited by more than one micro-RNA. This is because, unlike siRNAs, micro-RNAs have only partial complementarity to the target site. A description of micro-RNAs and their functioning can be found in [15–18]. An attempt to quantify the impact of each micro-RNA on the functioning of various genes is found in the program 'Targetscan', which is described in [19].

1.4 Role for the Systems and Control Community

In this section we present a broad philosophical discussion of how the systems and control community can contribute to cancer research. Some specific problems are discussed in Chap. 4 on future directions.

It is a truism that biology is in many ways far more complex than engineering. In engineering, one first designs a prototype system that performs satisfactorily, and then improves the design to be optimal (or nearly so), and finally, replicates the designed system as accurately as possible. In contrast, in biology, there is no standardization. Each of the 7 billion humans differ from each other in quite significant ways—clearly we are not mass-produced from a common template. Even if we focus on specific components of the human body and try to understand how they work together for a common purpose, there are difficulties. In designing complex engineering systems, each subsystem is designed separately, often by a dedicated design team. Then the subsystems are connected through appropriate isolators that ensure that, even after the various subsystems are interconnected, each subsystem still behaves as it was designed to. In contrast, in biology, it is difficult if not impossible to isolate individual subsystems and analyze their behavior. Even if one could succeed in understanding how a particular subsystem would behave in isolation, the behavior of the same subsystem gets altered significantly when it is a part of a larger system. This is one reason why many drugs that work well in vitro fail in vivo.

Because of these considerations, it is difficult for control theorists to make an impact on biology unless they work closely with experimental biologists. In a well-established subject like aerodynamics (to pick one), the fundamental principles are known and captured by the Navier-Stokes equation. Thus it is possible for an engineer to 'predict' how an airframe would behave to a very high degree of accuracy before metal is ever cut. In the author's view, given that in biology for the most part there are no foundational principles, and that measurement techniques are rather unreliable, control theorists must for the moment settle for a more modest role, namely 'generating plausible hypotheses' as opposed to 'making reliable predictions'. These plausible hypotheses are then validated or invalidated by experimentation. Learning is inductive: If a hypothesis is invalidated through experiment, then the model used to arrive at that hypothesis must be discarded; however, a confirmation of the hypothesis through experiment can serve only to increase one's confidence in the model.

In order to describe specific ways in which the controls community can contribute, we give a conceptual description of cancer therapy as it would appear to a control theorist. Because of the need to explain to a non-specialist readership, over-simplification is unavoidable, and the reader is cautioned that the description below is only 'probably approximately correct'. Those desirous of getting a more accurate picture should study the biology literature.

In the human body, cells die and are born all the time, and a rough parity is maintained between the two processes. Occasionally, in response to external stimuli, one or the other process gains the upper hand for a short period of time, and in a localized manner. For instance, if one gets a wound, then stem cells in the vicinity of the skin surrounding the wound proliferate rapidly for the production of new epidermis. Once the wound is closed, these cells return to a slower rate of replication. In the process of cell division and DNA replication, errors do occur. However, there is a fairly robust DNA repair process that corrects the errors made during replication. In spite of this, it is possible that some mutations that occurred during DNA replication do not get corrected, but instead get passed on to the next generation and the next after that; these are called somatic mutations. If these mutated cells replicate at a faster rate than normal cells, then it is possible (though not inevitable) that eventually the mutated cells overwhelm the normal cells by grabbing the resources needed for replication. At this point the cell growth, or tumor, has gone from being benign to being malignant. If the products of the mutated DNA enter the blood stream, or the lymph system, then the mutations can then be replicated at locations that are far-removed from the site of the original mutation; this is known as metastasis.

One of the complicating factors of cancer is that, in contrast to other diseases, every manifestation of the disease is in some sense unique.[5] Hence some sort of 'personal medicine' is not only desirable but imperative. Fortunately, thanks to all the advances cited in Sect. 1.3, there is now a tremendous opportunity to make personal medicine a reality. Specifically, by analyzing the vast amount of molecular and clinical data that is becoming available, cancer therapists can aspire to provide prognostic information and a selection of therapies that would most benefit each patient. In doing so, the following are some typical questions that need to be addressed:

- Given a genome-wide data set from a common experimental condition, is it possible to construct a context-specific (and genome-wide) gene interaction network (GIN) that describes how the various genes interact with each other in that particular experimental setting? Is it possible to validate a part of this network, at least around some genes (nodes)?
- Given a large number of patients with a particular form of cancer, is it possible to group them in such a way that the variation of their GINs within each group is a minimum, while at the same time the variation between groups is maximum?
- Using a combination of machine-learning (or statistical) and experimental methods, is it possible to predict which treatment regimen is likely to be most effective for a particular group of patients?

Within this broad framework, the control community can

- Integrate available data in a rational manner that would permit the generation of *all possible hypotheses that are consistent with the data* about therapeutic interventions.
- When the biologists come up with some hypotheses, exclude those hypotheses that are inconsistent with the data, and rank those that are consistent in terms of their statistical significance (i.e., compatibility with the available data).
- In a *suo motu* fashion, generate hypotheses that are suggested by the data, which the biologists can then validate.
- Recalibrate the statistical models to take into account new data as it becomes available, especially new data that does not match the predictions of existing models.

In short, by entering into a partnership with the biologists' community, the systems and control community can create a 'virtuous cycle' that would benefit both groups.

1.5 Organization of the Monograph

Due to space limitations, in this monograph we focus on one specific tool, namely the construction of context-specific genome-wide gene interaction networks (GINs) from expression data. The monograph is organized into four chapters including the

---

[5] One could paraphrase the opening sentence of Leo Tolstoy's *Anna Karenina* and say that 'Normal cells are all alike; every malignant cell is malignant in its own way'.

introduction. The contents of the three chapters after this one are briefly described next.

- Analyzing the statistical significance of labeled data. There are several methods that are currently used for this purpose, and the aim of this chapter is to gather in one place the most widely used techniques, and put them in context against each other.
- Reverse engineering context-specific genome-wide gene interaction networks (GINs) from expression data. As mentioned above, this is the main new research contribution of the monograph.
- Directions for new research. In this chapter we sketch some of the problems that can be tackled using the methods proposed here.

Throughout the monograph, we use a fairly homogenous set of ideas from probability and statistics, such as Markov chains, graphical models, goodness of fit tests etc. These should be familiar to most control theorists, so these concepts are just invoked as needed without further preamble. Those unfamiliar with these concepts can consult any standard reference.

References

1. Khammash, M., Tomlin, C., Vidyasagar, M.: Special issue on systems biology. IEEE Trans. Autom. Control IEEE Trans. Circ. Syst. Part I **53**(1), 1–241 (2008)
2. Allgöwer, F., Doyle, III, F.J.: Special issue on systems biology. Automatica (2011)
3. Mukherjee, S.: The Emperor of All Maladies. Fourth Estate, London (2011)
4. Mundi, A.: http://axismundi.spheresoflight.com.au
5. SEER: http://seer.cancer.gov/statfacts/html/all.html
6. Consortium, I.H.G.R.: Initial sequencing and analysis of the human genome. Nature **409**, 860–921 (2001)
7. Venter, J.C., Adams, M.D., Myers, E.W., et al.: The sequence of the human genome. Science **291**, 1304–1351 (2001)
8. TCGA: http://cancergenome.nih.gov
9. Babu, M.M.: An introduction to microarray data analysis. In: Computational Genomics: Theory and Application. Horizon Bioscience, Norwich (2004)
10. TCGA: The cancer genome atlas research network. Nature **474**, 609–615 (2011)
11. ChIP: http://en.wikipedia.org/wiki/chromatin_immunoprecipitation
12. ChIP-seq: http://ccg.vital-it.ch/chipseq/doc/chipseq_tutorial_intro.php
13. McLean, C.Y., et al.: Great improves functional interpretation of cis-regulatory regions. Nat. Biotechnol. **28**(5), 495–501 (2010)
14. Crick, F.H.C.: Central dogma of molecular biology. Nature **227**, 561–563 (1970)
15. Bartel, D.P.: Micrornas: genomics, biogenesis, mechanism, and function. Cell **116**(2), 281–297 (2004)
16. Lewis, B.P., CB, Burge, Bartel, D.P.: Conserved seed pairing, often flanked by adenosines, indicates that thousands of human genes are microrna targets. Cell **120**, 15–20 (2005)
17. Grimson, A., et al.: Microrna targeting specificity in mammals: determinants beyond seed pairing. Mol. Cell **27**, 91–105 (2007)
18. Bartel, D.: Micrornas: target recognition and regulatory functions. Cell **136**, 219–236 (2009)
19. Targetscan: http://www.targetscan.org

Chapter 2
Analyzing Statistical Significance

Abstract In this chapter we review some popular methods for estimating the statistical significance of various conclusions that can be drawn from experimental data. These include the χ^2-text, the Kolmogorov–Smirnov (K–S) test for goodness of fit, the 'student' t-test for testing the null hypothesis that two sets of data have the same mean, Significance Analysis for Microarrays (SAM), Pattern Analysis for Microarrays (PAM) and Gene Set Enhancement Analysis (GSEA).

Keywords Statistical significance tests · χ^2-test · Kolmogorov–Smirnov test · t-test · SAM · PAM · GSEA

In this chapter we will review some popular methods for estimating the statistical significance of various conclusions that can be drawn from experimental data. In a typical biological experiment, n distinct quantities called the 'features' are measured on each of m samples. The features can represent gene expression levels as measured in a micro-array study, levels of micro-RNA, or other things of this sort. Usually the true value of the feature is a nonnegative number. However, it is customary to normalize the data by centering (subtracting a bias term), scaling, and/or taking a logarithmic transform. Consequently the normalized feature values can also be negative numbers. The underlying assumption is that each feature $X_i, i = 1, \ldots, n$ is a random variable, and that for each sample index j, the vector $\mathbf{x}_j \in \mathbb{R}^n = (x_{ij}, i = 1, \ldots, n)$ is a measurement of the collection of random variables $\mathbf{X} = (X_i, i = 1, \ldots, n)$. It is further assumed that $\mathbf{x}_j$ is statistically independent of $\mathbf{x}_k$ for all $k \neq j$. However, there is no assumption that the X_i's themselves are independent random variables. Indeed determining the interdependence between the features is one of the key challenges in modeling, and is the subject of Chap. 3.

Thus the data set consists of an array of real numbers $\{x_{ij}\}, i = 1, \ldots, n, j = 1, \ldots m$. Moreover, it is often the case that the data is labeled. Thus the m samples are grouped into K classes, where class k consists of m_k samples (and obviously $\sum_{k=1}^{K} m_k = m$). Usually there is a biological basis for this grouping. For instance, K could equal two, and class 1 consists of measurements from tumor tissues of cancer

M. Vidyasagar, *Computational Cancer Biology*, SpringerBriefs in Control, Automation and Robotics, DOI: 10.1007/978-1-4471-4751-0_2, © The Author(s) 2012

patients who respond to a specific therapy, while class 2 consists of measurements from tumor tissues of cancer patients who are not responsive to the same therapy. Now all n features are measured on a new $(m + 1)$-st sample, and we would like to classify this new vector as belonging to one of the K classes. In order to do so, the type of questions that can be asked are the following:

- Suppose we divide the sample set into two classes consisting of m_1 and m_2 elements each, which without loss of generality can be renumbered as $\mathcal{M}_1 = \{1, \ldots, m_1\}$ and $\mathcal{M}_2 = \{m_1 + 1, \ldots, m_1 + m_2\}$ where $m_1 + m_2 = m$. For a specific feature (i.e., a specific index i), is it the case that average the expression level of feature i for class 1 differs at a statistically significant level from that of class 2? This is the topic of Sect. 2.1. How can this idea be extended to more than two classes?
- In biology it often happens that, in a collection of genes S (referred to as a genomic machine in Chap. 4), no single gene is over-expressed in class 1 compared to class 2; however, taken together they are over-expressed. Can this notion be made mathematically precise and tested? This is the subject of Sects. 2.2 and 2.3.
- Suppose some sort of classifier has been developed, which achieves a statistically significant separation between the various labeled classes. Now suppose, as before, that an $(m + 1)$-st data vector consisting of n expression level measurements becomes available. Usually such a classifier makes use of all n components of the data vector. Is it possible to identify a subset of $\{1, \ldots, n\}$ and a reduced-dimension classifier that more or less reproduces the classification abilities of a full-dimension classifier that uses all n components of the data? This is the topic of Sect. 2.4.

2.1 Basic Statistical Tests

In this subsection, we describe two basic tests, namely the t distribution and the Kolmogorov–Smirnov test for goodness of fit.

2.1.1 The t-Test

The 'student' t distribution can be used to test the null hypothesis that the means of two sets of samples are equal, under the assumption that the variance of the two sample sets is the same. If the variance of the two sample sets is not the same (or is not presumed to be the same), then the 'student' t-test is replaced by Welch's t-test. Strictly speaking the t distribution is derived for the case where the samples follow a normal distribution. However, it can be shown that the distribution applies to a wide variety of situations, even without the normality assumption.

Let us begin with the 'student' t-test, named after William Seely Gossett, who first published this test anonymously under the name 'Student'. Suppose we have two classes of samples $\mathcal{M}_1$, $\mathcal{M}_2$, of sizes m_1, m_2 respectively. Thus the data consists of $x_1, \ldots, x_{m_1}$ belonging to the class $\mathcal{M}_1$, and $x_{m_1+1}, \ldots, x_{m_1+m_2}$ belonging to the class $\mathcal{M}_2$. Let

$$\bar{x}_i = \frac{1}{m_i} \sum_{j \in \mathcal{M}_i} x_j, i = 1, 2,$$

denote the means of the two sample classes. Then it is well-known that $\bar{x}_i$ is an unbiased estimate of the true but unknown expected value of the samples in class i. Next, let us define

$$S_i^2 = \frac{1}{m_i - 1} \sum_{j \in \mathcal{M}_j} (x_j - \bar{x}_i)^2, i = 1, 2.$$

Then it is again well-known that S_i^2 is an unbiased estimate of the variance of the samples in class i. Since it is being assumed that both classes have the same variance, these two estimates are 'pooled' to obtain an estimate of the variance of the overall samples, as follows:

$$S_P^2 = \frac{(m_1 - 1)S_1^2 + (m_2 - 1)S_2^2}{m_1 + m_2 - 2}$$

$$= \frac{1}{m_1 + m_2 - 2} \sum_{i=1}^{2} \sum_{j \in \mathcal{M}_j} (x_j - \bar{x}_i)^2. \tag{2.1}$$

In other words, the pooled variance is just a weighted average of the two unbiased variance estimates of each class.

The 'student' or Gossett version of the t-test consists of the observation that the test statistic

$$d_t = \frac{\bar{x}_1 - \bar{x}_2}{S_P \sqrt{(1/m_1) + (1/m_2)}} \tag{2.2}$$

satisfies the t distribution with $m_1 + m_2 - 2$ degrees of freedom. Note that as the number of degrees of freedom approaches infinity, the t distribution approaches the normal distribution. In practice, the t distribution is virtually indistinguishable from the normal distribution when the number of degrees of freedom becomes 20 or larger. Explicit but complicated formulae are available in the literature for the probability density and cumulative distribution function of the t distribution. Note that this test is also sometimes referred to as the 'two-sample' t-test, because we are comparing two empirically determined means against each other. The 'one-sample' t-test would be to test whether or the true but unknown mean of a given data set equals a prespecified number.

In case it is *not* assumed that the variances of the two samples are equal, then the student t-test is replaced by Welch's t-test. In this case, the 'pooled' variance estimate defined in (2.1) is replaced by

$$S_W = \sqrt{\frac{S_1^2}{m_1} + \frac{S_2^2}{m_2}}, \tag{2.3}$$

while the number of degrees of freedom is now given by

$$dofw = \frac{(S_1^2/m_1 + S_2^2/m_2)^2}{(S_1^2/m_1)^2/(m_1 - 1) + (S_2^2/m_2)^2/(m_2 - 1)}.$$

Note that, in contrast with the student t-test, in the case of Welch's test the number of degrees of freedom need not be an integer, in which case it is truncated to the nearest integer. The Welch t-test consists of the observation that the quantity

$$d_W = \frac{\bar{x}_1 - \bar{x}_2}{S_W} \tag{2.4}$$

satisfies the t distribution with $\lfloor dofw \rfloor$ degrees of freedom.

The t test is applied as follows: Given the two sets of samples, the null hypothesis is that their means are the same. If the assumption is that the variances of the two samples are the same, then the test statistic d_t is computed from (2.2) for the actual samples. Using the standard tables, the likelihood that a random variable X with the t distribution exceeds d_t (if $d_t > 0$) or is less than d_t (if $d_t < 0$) is computed. If this likelihood is smaller than some prespecified level δ, then the null hypothesis is rejected at the level δ. In other words, it can be concluded with confidence $1 - \delta$ that the null hypothesis is false. In case it is not assumed that the variances are the same, the above procedure is applied with d_t replaced by d_W.

2.1.2 The Kolmogorov–Smirnov (K–S) Tests

Next we describe the Kolmogorov–Smirnov (K–S) tests for goodness of fit. Suppose X is a real-valued random variable (r.v.). Then its **cumulative distribution function (cdf)**, denoted by $\Phi_X(\cdot)$, is defined by

$$\Phi_X(u) = \Pr\{X \leq u\},$$

while the **complementary distribution function**, denoted by $\bar{\Phi}_X(\cdot)$, is defined by

$$\bar{\Phi}(u) = 1 - \Phi_X(u) = \Pr\{X > u\}.$$

The cdf of any r.v. has a property usually described as 'cadlag', which is an acronym formed from the French phrase 'continu à droite, limité à gauche'. In other words, the cdf is right-continuous in the sense that

$$\lim_{u \to u_0^+} \Phi_X(u) = \Phi_X(u_0),$$

and it has left limits in the sense that the limit

$$\lim_{u \to u_0^-} \Phi_X(u) =: \Phi_X^-(u_0)$$

exists and satisfies $\Phi_X^-(u_0) \leq \Phi_X(u_0)$ for all real u_0.

Suppose $\mathbf{x} = \{x_t\}_{t \geq 1}$ are independent samples of X. Based on the first l samples, we can construct an 'empirical cdf' of X, as follows:

$$\hat{\Phi}_l(u) := \frac{1}{l} \sum_{i=1}^{l} I_{\{x_i \leq u\}}, \tag{2.5}$$

where I is the indicator function; thus I equals one if the condition stated in the subscript is true, and equals 0 if the condition stated in the subscript is false. To put it another way, $\hat{\Phi}_l(u)$ is just the fraction of the first l samples that are less than or equal to u. The quantity

$$D_l := \sup_u |\hat{\Phi}_l(u) - \Phi_X(u)|$$

gives a measure of just how well the empirical cdf approximates the true cdf. The well-known Glivenko-Cantelli lemma [1, 2], [3, p. 448], [4, p. 20] states that the stochastic process $\{D_l\}$ converges almost surely to zero as $l \to \infty$.

In the case where the true but unknown cdf $\Phi_X(\cdot)$ is continuous, the theorems of Kolmogorov [5] and Smirnov [6] quantify the *rate of convergence*, thereby leading to a test for goodness of fit.[1] See also [7] for simpler proofs of these two theorems. Specifically, let us think of D_l as a real-valued random variable, and let Φ_{D_l} denote the cdf of D_l. Then Kolmogorov [5] has shown that, for every fixed $u > 0$,

$$\Phi_{D_l}(u) \to \Phi_K(u\sqrt{l}) \text{ as } l \to \infty, \ \bar{\Phi}_{D_l}(u) \to \bar{\Phi}_K(u\sqrt{l}) \text{ as } l \to \infty, \tag{2.6}$$

where Φ_K is the Kolmogorov cdf given by

$$\Phi_K(u) = 1 - 2\sum_{k=1}^{\infty} (-1)^{k+1} \exp(-2k^2 u^2), \tag{2.7}$$

$$\bar{\Phi}_K(u) = 2\sum_{k=1}^{\infty} (-1)^{k+1} \exp(-2k^2 u^2). \tag{2.8}$$

---

[1] Note that the Glivenko-Cantelli lemma does not require $\Phi_X(\cdot)$ to be continuous.

and the convergence is in the distributional sense. Recall that a sequence of random variables $\{Y_l\}$ converges to another random variable Z in the distributional sense if

$$\sup_u |\Phi_{Y_l}(u) - \Phi_Z(u)| \to 0 \text{ as } l \to \infty.$$

Thus the contribution of Kolmogorov lies in determining the exact limit of the cdf of the error term D_l. Observe from (2.6) that, no matter how small u is, provided only that $u > 0$, the quantity $u\sqrt{l} \to \infty$ as $l \to \infty$, whence $\bar{\Phi}_{D_l}(u) \to 0$ as $l \to \infty$.

The Kolmogorov test is used to validate the null hypothesis that a given set of samples $x_1, \ldots, x_l$ are generated in an i.i.d. fashion from a specified cdf $F(\cdot)$. To apply the test, we first construct the empirical cdf $\hat{\Phi}_l$ as in (2.5), and then compute the goodness of fit statistic

$$d_l = \sup_u |\hat{\Phi}_l(u) - F(u)|.$$

Then the null hypothesis is rejected at level δ (that is, with confidence $\geq 1 - \delta$) if

$$\sqrt{l}d_l > (\bar{\Phi}_K)^{-1}(\delta),$$

where $\bar{\Phi}_K(u) = 1 - \Phi_K(u)$ is the complementary distribution function. This is called the one-sample K-S test, though strictly speaking it should really be called the Kolmogorov test.

Smirnov [6] extended the Kolmogorov test to the case where there are two sets of samples $x_1, \ldots, x_l$ and $y_1, \ldots, y_m$, possibly of different lengths. The null hypothesis is that both sets of samples are generated from a common, but unspecified (and continuous), cdf. To test this hypothesis, we form two empirical cdfs, call them $\hat{\Phi}_l$ based on the x_i samples, and $\hat{\Psi}_m$ based on the y_j samples, in analogy with (2.5). Smirnov's theorem is that if we define the random variable

$$D_{l,m} = \sup_u |\hat{\Phi}_l(u) - \hat{\Psi}_m(u)|$$

and

$$n = \frac{lm}{l+m} = \left(\frac{1}{l} + \frac{1}{m}\right)^{-1},$$

then

$$\Phi_{D_{l,m}}(u) \to \Phi_K(u\sqrt{n}) \text{ as } \min\{l, m\} \to \infty, \tag{2.9}$$

$$\bar{\Phi}_{D_{l,m}}(u) \to \bar{\Phi}_K(u\sqrt{n}) \text{ as } \min\{l, m\} \to \infty. \tag{2.10}$$

This is sometimes (erroneously) called the two-sample K-S test. To apply this test, one first computes the test statistic

$$d_{l,m} = \sup_u |\hat{\Phi}_l(u) - \hat{\Psi}_m(u)|.$$

The null hypothesis, namely that both sets of samples are coming from the same (but unknown) cdf, is rejected at level δ if

$$\sqrt{\frac{lm}{l+m}} d_{l,m} > (\bar{\Phi}_K)^{-1}(\delta).$$

Recall that

$$\Phi_K(u) = 1 - 2\sum_{k=1}^{\infty}(-1)^{k+1}\exp(-2k^2u^2),$$

$$\bar{\Phi}_K(u) = 2\sum_{k=1}^{\infty}(-1)^{k+1}\exp(-2k^2u^2).$$

Though the above formulas (2.6) and (2.10) are explicit, it is very difficult to compute $(\bar{\Phi}_K)^{-1}(\delta)$ for a given number δ. However, if we are willing to forgo a little precision, a simple estimate can be derived. Observe that $\bar{\Phi}_K(u)$ is defined by an alternating series; as a result $\bar{\Phi}_K(u)$ is bracketed by any two successive partial sums. In particular, we have that

$$\bar{\Phi}_K(u) \leq 2\exp(-2u^2) =: \bar{\Phi}_M(u), \quad \forall u.$$

Therefore it follows that

$$(\bar{\Phi}_K)^{-1}(\delta) \leq (\bar{\Phi}_M)^{-1}(\delta), \quad \forall \delta.$$

So to apply the one-sample K-S test, we reject the null hypothesis at level δ if

$$\sqrt{l}d_l > (\bar{\Phi}_M)^{-1}(\delta) \iff \bar{\Phi}_M(\sqrt{l}d_l) < \delta$$

$$\iff 2\exp(-2ld_l^2) < \delta$$

$$\iff d_l > \left[\frac{1}{2l}\log\frac{2}{\delta}\right]^{1/2}.$$

Let us define

$$\theta_M(l,\delta) := \left[\frac{1}{2l}\log\frac{2}{\delta}\right]^{1/2} \tag{2.11}$$

to be the K-S threshold as a function of the number of samples l and the level δ. With this notation, the null hypothesis is rejected at level δ if d_l exceeds this threshold. Note that the above threshold is 'conservative' because we have replaced the exact value $\bar{\Phi}_K^{-1}(\delta)$ by its upper bound $\bar{\Phi}_M^{-1}(\delta)$. But in return we have a very explicit formula for the threshold.

Now we digress briefly to discuss how the above kind of tests can be applied in more general contexts. As stated, the K-S tests apply strictly to real-valued random variables, and that too, only when the cdf of the underlying random variables is continuous. Extending it even to r.v.s assuming values in $\mathbb{R}^d$ when $d \geq 2$ is not straight-forward; see [8] for one of the few results in this direction. The objective of this digression is to point out that, if one were to use recent results in statistical learning, then K-S-like tests are abundant in quite general settings. A good reference for the discussion below is [9].

We begin with the observation that the 'modern' way to prove the Glivenko-Cantelli lemma is to apply Vapnik-Chervonenkis, or VC theory, and sketch the main results of the theory next. Suppose X is some set (which need not be a subset of a Euclidean space such as $\mathbb{R}^d$), and that P is a probability measure on X. Suppose i.i.d. samples $\{x_t\}_{t \geq 1}$ are generated from X according to the law P. Let $\mathcal{A}$ denote some collection of subsets of X.[2] For each set $A \in \mathcal{A}$, we compute an empirical probability

$$\hat{P}_l(A) = \frac{1}{l} \sum_{t=1}^{l} I_{\{x_i \in A\}}.$$

In other words, $\hat{P}_l(A)$ is just the fraction of the l samples that belong to the set A. Finally, in analogy with earlier notation, define

$$D_l := \sup_{A \in \mathcal{A}} |\hat{P}_l(A) - P(A)|.$$

The collection of sets $\mathcal{A}$ has the property of 'uniform convergence of empirical means' if $D_l \to 0$ almost surely as $l \to \infty$.

Recent developments in statistical learning theory, specifically VC theory, consist of associating with each collection of sets $\mathcal{A}$ a positive integer d, called the VC-dimension of $\mathcal{A}$. One of the main results of this theory as described in [9, Theorem 7.4] states that if d is finite, then the collection does indeed has the uniform convergence property. Moreover, if $\bar{\Phi}_{D_l}$ denotes the complementary df of the random variable D_l, then it can be stated with confidence $1 - \delta$ that

$$\bar{\Phi}_{D_l}(u) \leq 4 \left(\frac{2el}{d} \right)^d \exp(-lu^2/8), \tag{2.12}$$

where e denotes the base of the natural logarithm. In particular, the collection of semi-infinite intervals $\{(-\infty, u], u \in \mathbb{R}\}$ has VC-dimension one, so that for the standard

---

[2] Strictly speaking, we should first define a σ-algebra $\mathcal{S}$ of subsets of X and assume that $\mathcal{A} \subseteq \mathcal{S}$. Such details are glossed over here but the treatment in [9] is quite precise.

K-S setting, we can state with confidence $1 - \delta$ that

$$\bar{\Phi}_{D_l}(u) \leq 8el \exp(-lu^2/8).$$

In higher dimensions, say in $\mathbb{R}^d$, the collection of sets

$$\mathcal{A} = \{\prod_{i=1}^{d}(-\infty, u_i], u_i \in \mathbb{R} \; \forall i\}$$

has VC-dimension equal to d, so that (2.12) holds.

To apply this bound in a general setting, suppose P is some probability measure on X, and that $x_1, \ldots, x_l$ are elements of X. The null hypothesis is that these samples have been generated as independent samples according to the law P. To test this hypothesis, choose any collection of subsets $\mathcal{A}$ of X with finite VC-dimension d, and form the test statistic

$$d_l = \sup_{A \in \mathcal{A}} |P(A) - \hat{P}_l(A)|.$$

If it is the case that $\bar{\Phi}_{D_l}(d_l) \leq \delta$, then the null hypothesis is rejected the level δ. Now we don't know $\bar{\Phi}_{D_l}(d_l)$ but we do have an upper bound in the form of (2.12). Let $\bar{\Phi}_{VC}$ denote the right side of (2.12). Then we reject the null hypothesis at level δ if $\bar{\Phi}_{VC}(d_l) \leq \delta$. This can be turned into an explicit threshold formula by simple algebra. It is easy to show that

$$\bar{\Phi}_{VC}(d_l) \leq \delta \iff d_l \geq \left[\frac{8}{l}\left(\log\frac{4}{\delta} + d\log\frac{2el}{d}\right)\right]^{1/2}.$$

Let us denote the right side as a new threshold function, namely

$$\theta_{VC}(l, \delta; d) := \left[\frac{8}{l}\left(\log\frac{4}{\delta} + d\log\frac{2el}{d}\right)\right]^{1/2}. \tag{2.13}$$

Then we reject the null hypothesis if $d_l > \theta_{VC}(l, \delta; d)$.

If we compare the thresholds from K-S theory and VC theory, we see from (2.11) and (2.13) that for fixed confidence level δ the K-S threshold is $O(l^{-1/2})$ whereas the VC threshold is $O(l^{-1/2} \log l)$. But the VC threshold is far more general. So the slightly more conservative bound is definitely worthwhile. For fixed sample length l, both thresholds are $O(\log(1/\delta))$ so there is no difference.

2.2 Significance Analysis for Microarrays

In this subsection we discuss a widely use method called Significance Analysis for Microarrays (SAM), introduced in [10]. The reader is directed to that paper for discussion of earlier work in this area.

The problem considered is the following: Suppose as before that we have a gene expression data set $\{x_{ij}\}$, $i = 1, \ldots, n$, $j = 1, \ldots, m$, where n is the number of genes and m is the number of samples. Suppose further that the data is labeled and divided into two classes. Without loss of generality, suppose the first m_1 samples belong to class 1, and the remaining $m_2 = m - m_1$ belong to class 2. We would like to assess which amongst the n genes show significant variation between the two classes.

As a first-cut, we could treat each of the n genes separately, and for each index i, construct a two-sample t-test statistic between the samples $\{x_{ij}, j = 1, \ldots, m_1\}$ and $\{x_{ij}, j = m_1+1, m_1+m_2\}$. Specifically, for each index i, let $\bar{x}_{i1}, \bar{x}_{i2}$ denote the average values of the samples in the two classes, and the pooled standard deviation S_i by

$$S_i^2 = \frac{1}{m_1 + m_2 - 2} \left[\sum_{j=1}^{m_1} (x_{ij} - \bar{x}_{i1})^2 + \sum_{j=m_1+1}^{m} (x_{ij} - \bar{x}_{i2})^2 \right].$$

Now it can happen that some genes exhibit so little variation within each class that S_i is very small, with the consequence that any quantity divided by S_i automatically becomes large. To guard against this possibility, a constant S_0 is chosen to be the same for all indices i, and is added to S_i. Next, for each index i, we define the test statistic

$$\alpha_{i0} = \frac{\bar{x}_{i1} - \bar{x}_{i2}}{(S_i + S_0)[(1/m_1) + (1/m_2)]^{1/2}}.$$

By examining the significance of α_{i0} using the t-distribution and the two-sample t-test, we might be able to determine whether gene i exhibits a substantial variation between the two classes.

However, this alone might not give a true picture. It often happens in the case of biological data that the *inherent* variation of expression levels changes enormously from one gene to another. For instance, the expression level of one gene may show barely 10% variation across experiments, whereas that of another gene may show an order of magnitude variation. If we were to apply the K-S test blindly, we would conclude that the second gene is far more significant than the first one. But this is potentially misleading. In biology it is often the case that the downstream consequences of variations in gene expression are also widely different for different genes.

To normalize against this possibility, in [10], the authors introduce an additional criterion. Given the integers m_1, m_2, choose an integer k roughly equal to

$0.5 \min\{m_1, m_2\}$. Let $\pi_1, \ldots, \pi_L$ be permutations of $\{1, \ldots, m\}$ into itself such that precisely k elements from class 1 are shifted to class 2 and vice versa. In the original paper [10] $m_1 = m_2 = 4$ so that $k = 2$, and there are $6^2 = 36$ such permutations; so they consider all of them. However, if the integers m_1, m_2 are sufficiently large, the number of such permutations will be huge, in which case one chooses, at random, a prespecified number L of such permutations. For each permutation π_l, the first m_1 elements are labeled as 1 and the rest are labeled as 2. In other words, the elements $\pi_l(1), \ldots, \pi_l(m_1)$ are given the label 1 while the rest are given the label 2. For each labeling corresponding to the permutation π_l, let us compute a two-sample t-test statistic, which we may denote by α_{il}. This is done for each of the n genes. Next, let us define

$$\alpha_E(i) = \frac{1}{L} \sum_{l=1}^{L} \alpha_{il}$$

to be the value of the test statistic averaged across all L permutations. Let α_{i0} denote the test statistic corresponding to the identity permutation, that is, the original labeling. For most genes (i.e., for most indices i), the test statistic α_{i0} corresponding to the original labeling will not differ much from the averaged value $\alpha_E(i)$. Those genes for which the difference is significant, in either direction, are the genes that one should examine. To implement this criterion, an absolute constant Δ is chosen, and only those genes for which $|\alpha_{i0} - \alpha_E(i)| \geq \Delta$ are studied further. One could of course argue that the threshold should be in terms of the ratio $\alpha_{i0}/\alpha_E(i)$ and that too would be a valid viewpoint. In [10], using this approach only 46 out of an original set of 6,800 genes are found to be worth examining further—a reduction of more than two orders of magnitude. What this means is that, for all except these 46 genes, the test statistic corresponding to the original labeling is not very different from what would result from a purely random assignment of labels. These short-listed genes are then examined whether indeed there is substantial variation between the two classes (which it may be noted is a different question from whether a randomly assigned label would result in a different value for the test statistic). A gene belonging to this shorter list is deemed to exhibit significant variation between classes 1 and 2 if

$$\max\left\{ \frac{\bar{x}_{i1}}{\bar{x}_{i2}}, \frac{\bar{x}_{i2}}{\bar{x}_{i1}} \right\} > R,$$

where R is another threshold. This thresholding results in a final set of genes with two attributes: (i) The test statistic corresponding to the original labeling differs substantially from that corresponding to a random assignment of labels, and (ii) there is substantial difference between the mean values of the two classes. This is the desired list of genes. Note that we could have just as easily compared $|\log(\bar{x}_{i1}/\bar{x}_{i2})|$ against a threshold. We could also apply the K-S test and choose those genes for which the difference is statistically significant at a prespecified level.

2.3 Gene Set Enhancement Analysis

As in the previous subsection, suppose have a gene expression data set $\{x_{ij}\}$, $i = 1, \ldots, n, j = 1, \ldots, m$, where n is the number of genes and m is the number of samples. Further, the data is labeled and divided into two samples. Suppose $\mathcal{M} = \{1, \ldots, m\}$ and that $\mathcal{M}_1, \mathcal{M}_2$ is a partition of $\mathcal{M}$. Further, suppose $|\mathcal{M}_i| = m_i$ for $i = 1, 2$. For example, the samples in class 1 may come from healthy tissue while those in class 2 may come from cancerous tissue. In the previous subsection, we studied the problem of identifying *individual genes* within the set of n genes that show statistically significant variation between the two classes. For this purpose, for each gene i we compared the t-statistic between the two classes against what would be obtained by randomly assigning labels to the of m samples associated with that gene. In this section, we carry the discussion to a greater level of generality. Specifically, it can happen in biological experiments that, while no single gene may show statistically a significant difference between the two classes, a collection of genes acting in concert may exhibit such statistically significant difference between the two classes. Accordingly, suppose a subset S of $\mathcal{N} = \{1, \ldots, n\}$ is specified beforehand as a set of genes that we expect might *collectively* exhibit different expression levels between the two classes. Note that the set S is specified on the basis of biological considerations, and not deduced *post facto* from the data under study. For instance, S could be one of the 'genomic machines' identified through the Netwalk algorithm of Sect. 4.2.

The discussion below is essentially taken from [11] which describes an algorithm that those authors call Gene Set Analysis (GSA). In turn [11] builds on an earlier algorithm called Gene Set Enhancement Analysis (GSEA) from [12]. Along the way, the authors of [11] also relate their GSA algorithm to several earlier algorithms. In the interests of conserving space, we do not reference nor discuss all the earlier work, and the interested reader is directed to the bibliography of [11].

The main idea of the GSA algorithm is the following: In SAM (Significance Analysis for Microarrays) discussed in Sect. 2.2, for each index i denoting the gene, we did the following: First we computed the t-statistic of the difference between the means of the two classes. Then we assigned random labels to the m samples associated with gene i, ensuring that m_i are placed in class i, and for each random labeling, we computed the same t-statistic. That is fine so far as testing a single gene goes. To test whether a prespecified set of genes S shows significant difference between the two classes, it is necessary to perform an additional step, as described next. Let $k = |S|$. Then, in addition to permuting the labels of the m columns associated with each gene in the set $\mathcal{N}$, we should also do the same to a randomly selected set of k genes from the collection $\mathcal{N}$. In [11], assigning the class labels at random is referred to as 'permutation' while choosing a random set of k genes from $\mathcal{N}$ is referred to as 'randomization'. An additional complication in the randomization step is that, while the expression levels of k randomly selected genes from $\mathcal{N}$ can be thought of as being uncorrelated, the expression levels of the k genes in the specified set S are quite likely to be correlated (due to their having a common biological function etc.).

Hence the randomized data will in general have different statistical behavior from that of the genes in the set S. The GSA algorithm attempts to correct for this feature.

The details of the algorithm are as follows: For each gene i in $\mathcal{N}$, form a two-sample t-statistic, call it d_i. Then d_i is distributed according to the t-distribution with $m - 2$ degrees of freedom. The quantity d_i is transformed into another value z_i that has a normal distribution, by the rule

$$z_i = \Phi_{\mathrm{Nor}}^{-1}(\Phi_{t,m-2}(d_i)),$$

where Φ_{Nor} denotes the cdf of a normal r.v. and $\Phi_{t,m-2}$ denotes the cdf of a t-distributed r.v. with $m - 2$ degrees of freedom. Note that if the number of samples m is sufficiently large, then the t-distribution is virtually identical to the normal distribution, so this step can be omitted. Now suppose $S : \mathbb{R} \to \mathbb{R}$ is a scoring function.[3] In [12], the scoring function $S(z)$ equals $|z|$. For each gene i, let S_i be a shorthand for $S(z_i)$. For the gene set S, compute the score

$$S = \frac{1}{k} \sum_{i \in S} S_i. \tag{2.14}$$

The question under study is: Is the score S sufficiently significant?

Now compute the mean μ_0 and standard deviation σ_0 of the raw samples in the familiar manner, namely:

$$\mu_0 = \frac{1}{n} \sum_{i \in \mathcal{N}} S_i, \ \sigma^2 = \frac{1}{n-1} \sum_{i \in \mathcal{N}} (S_i - \mu_0)^2.$$

Next, choose at random several subsets of $\mathcal{N}$ of cardinality k, compute the counterpart of the score S for each such randomly chosen gene set, and compute the mean and standard deviation of all of these scores (over all the randomly selected sets of cardinality k). Denote these by $\mu^{\dagger}, \sigma^{\dagger}$ respectively. If all the samples S_i within a set of cardinality k are independent, then we would have $\mu^{\dagger} = \mu_0, \sigma^{\dagger} = \sigma/\sqrt{k}$. But this need not be the case in general.

Next, choose a large number of permutations $\pi_1, \ldots, \pi_L$ of $\mathcal{M}$ into itself. For each permutation π_l, assign the label i to the samples in the image $\pi_l(\mathcal{M}_i)$, for $i = 1, 2$. This will generate, for each gene i, a test statistic $z_{\pi_l,i}$ and score $S_{\pi_l,i}$. Let μ_P, σ_P denote the mean and standard deviation of these nL numbers, where the subscript P is to remind us of 'permutation'.

The next step is called 'restandardization'. For each permutation π_l, let S_{π_l} denote the score resulting from the labeling as per the permutation π_l. Then the renormalized score corresponding to π_l is defined as

---

[3] We mostly follow the notation in [11], in which the letter S in various fonts is used to denote various quantities. The reader is therefore urged to pay careful attention.

$$S_{R,\pi_l} = \mu^{\dagger} + \frac{\sigma^{\dagger}}{\sigma_P}(S_{\pi_l} - \mu_P).$$

Then a test statistic is given by the quantity

$$p_S = \frac{1}{L} \sum_{l=1}^{L} I_{\{S_{R,\pi_l} > S\}},$$

which is the fraction of the restandardized scores that exceed the nominal score S. Clearly the smaller p_S is, the more significant is the score S. In GSEA, the cdf of the samples $\{z_i, i \in S\}$ is compared to the cdf of the samples $\{z_i, i \notin S\}$. This more or less corresponds to the choice $s(z) = |z|$.

Finally, in [11] another statistic is introduced, known as the max-mean statistic. Define

$$(z)_+ = \max\{z, 0\}, (z)_- = -\min\{z, 0\},$$

and observe that $(z)_-$ is positive if z is negative, somewhat contrary to the usual convention. Now define

$$s^+ = \frac{1}{k} \sum_{i \in S}(s_i)_+, s^- = \frac{1}{k} \sum_{i \in S}(s_i)_-, s_{\max} = \max\{s^+, s^-\}.$$

2.4 Pattern Analysis for Microarrays

In this subsection we discuss a method for simplifying the application of nearest neighbor clustering in the context of gene expression studies. This method is known as Pattern Analysis for Microarrays (PAM) [13]. The similarity of the acronyms SAM and PAM is not coincidental, because as we shall see, the two approaches have a lot in common.

As always, suppose we are given a set of gene expression data $\{x_{ij}\}$, $i = 1, \ldots, n, j = 1, \ldots, m$. Suppose further that the set $\mathcal{M} = \{1, \ldots, m\}$ of samples is divided into K classes, which are denoted here as $\mathcal{M}_k, k = 1, \ldots, K$. Thus the collection $\{\mathcal{M}_1, \ldots, \mathcal{M}_K\}$ is a partition of $\mathcal{M}$. Let denote $|\mathcal{M}_k|$ by m_k. Now suppose a new data vector $\mathbf{y} \in \mathbb{R}^n$ arrives from a fresh study. We would like to classify $\mathbf{y}$ as belonging to one of the K classes. How should we go about it?

One of the most commonly used method is that of nearest neighbor classification. As before, let us define the mean values of the expression level of gene i in class k, and the overall mean value, by

$$\bar{x}_{ik} := \frac{1}{m_k} \sum_{j \in \mathcal{M}_k} x_{ij}, k = 1, \ldots, K,$$

$$\bar{x}_i = \frac{1}{m} \sum_{k=1}^{K} \sum_{j \in \mathcal{M}_k} x_{ij} = \sum_{k=1}^{K} \frac{m_k}{m} \bar{x}_{ik}.$$

Thus $\bar{\mathbf{x}}_k \in \mathbb{R}^n$ is the centroid of class k while $\mathbf{x} \in \mathbb{R}^n$ is the overall centroid. To classify the vector $\mathbf{y}$, we compute the Euclidean distance to each of the K centroids, and classify it into the class whose centroid is the closest. Applying this classification method requires the computation of

$$\|\mathbf{y} - \bar{\mathbf{x}}_k\|^2 = \sum_{i=1}^{n} (y_i - \bar{x}_{ik})^2 \qquad (2.15)$$

for each k. If, as is often the case, n is of the order of thousands if not tens of thousands, the above computation can be quite expensive. The objective of PAM is to determine a subset $\mathcal{N}_1$ of $\mathcal{N} = \{1, \ldots, n\}$ with $|\mathcal{N}_1| \ll n$ such that, if the summation is taken only over those $i \in \mathcal{N}_1$, the resulting nearest neighbor classification would be more or less the same.

The basic idea behind PAM is as follows: In [13], PAM is also referred to as the 'method of shrunken centroids'. Suppose that for some index i, it is the case that $\bar{x}_{ik}$ is the same for all values of k. In other words, suppose that the i-th component of the centroid $\bar{\mathbf{x}}_k$ is the same for all k. Then it is obvious that the index i can be dropped from the summation in (2.15) because the term $(y_i - \bar{x}_{ik})^2$ makes an equal contribution for all k. So the method of shrunken centroids consists of shrinking the spread amongst $\{x_{i1}, \ldots, x_{iK}\}$ to zero for as many indices i as possible, by replacing the true centroid by a synthetic centroid.

In analogy with earlier reasoning, define the pooled within class standard deviation of gene i by

$$S_i^2 = \frac{1}{m - k} \sum_{k=1}^{K} \sum_{j \in \mathcal{M}_k} (x_{ij} - \bar{x}_{ik})^2.$$

Next, as before, a small constant S_0 (independent of i) is added to each S_i to avoid division by very small numbers. Now define a test statistic d_{ik} that tests for the null hypothesis that the data in class k differs significantly from the overall data, namely

$$d_{ik} = \frac{\bar{x}_{ik} - \bar{x}_i}{(S_i + S_0)[(1/m_k) + (1/m)]^{1/2}} =: \frac{\bar{x}_{ik} - \bar{x}_i}{l_k(S_i + S_0)},$$

where

$$l_k = \left[\frac{1}{m_k} + \frac{1}{m} \right]^{1/2}.$$

Note that it would perhaps be more accurate to compare $\bar{x}_{ik}$ with the 'leave one out' mean of all the remaining $m - m_k$ entries, as opposed to the overall mean $\bar{x}_i$. But this would involve considerably more computation with relatively little benefit.

Now rewrite the above relationship as

$$\bar{x}_{ik} = \bar{x}_i + l_k(S_i + S_0)d_{ik}.$$

If we could somehow justify replacing the actual d_{ik} by zero, then it would follow that $\bar{x}_{ik} = \bar{x}_i$ for all k, and we could therefore ignore the i-th term in the summation (2.15). This is achieved by soft thresholding. Specifically, a fixed constant Δ, independent of both i and k, is selected, Then we define

$$d'_{ik} = \text{sign}(d_{ik})(|d_{ik}| - \Delta)_+,$$

where as usual $(x)_+ = \max\{x, 0\}$. An equivalent definition of d'_{ik} is

$$d'_{ik} = \begin{cases} d_{ik} - \Delta, & \text{if } d_{ik} > \Delta, \\ d_{ik} + \Delta, & \text{if } d_{ik} < -\Delta, \\ 0, & \text{if } |d_{ik}| \le \Delta. \end{cases}$$

Then the centroids are 'shrunk' by replacing d_{ik} by d'_{ik}, namely

$$\bar{x}_{ik} = \bar{x}_i + l_k(s_i + s_0)d'_{ik}. \tag{2.16}$$

Note that if $d'_{ik} = 0$ for all k for a fixed i, then that term can be dropped from the summation in (2.15).

The higher the value of Δ, the more thresholds that will be set to zero. At the same time, the higher the value of Δ, the more the likelihood of misclassification by the simplified summation. In [13], the constant Δ is chosen through ten-fold cross validation. The data set is divided vertically (in terms of the index j) into ten more or less equal-sized data sets. Then 90% of the data is used as training data and the remaining 10% is used to test the resulting reduced-sum classifier; this exercise is repeated by shifting the testing data through each subset of the data. The constant Δ is adjusted up or down until the cross-validation produces satisfactory results. In [13], the original data set consists of expression levels of 2,308 genes, 63 samples, classified into four forms of cancer. Thus $n = 2308$, $m = 63$ and $K = 4$. By using the soft thresholding technique, a subset of a mere 43 'most useful genes' are identified out of the original 2,308—a reduction of about 98% in the computational burden.

References

1. Glivenko, V.I.: Sulla determinazione empirica delle legge di probabilità. Giorn. dell'Ist. Italia degli Attuari **4**, 92–99 (1933)
2. Cantelli, F.P.: Sulla determinazione empirica delle legge di probabilità. Giorn. dell'Ist. Italia degli Attuari **4**, 421–424 (1933)
3. Gnedenko, B.V.: Theory of probability, 4th edn. Chelsea, New York (1968)
4. Loève, M.: Probability Theory I, 4th edn. Springer, Heidelberg (1977)

5. Kolmogoroff, A.: Sulla determinazione empirica di una legge di probabilita. Giorn. dell'Ist. Italia degli Attuari **4**, 83–91 (1933)
6. Smirnov, N.: Ob uknonenijah empiriceskoi krivoi raspredelenija. Recueil Mathematique (Matematiceskii Sbornik) **6**(48), 3–26 (1939)
7. Feller, W.: On the kolmogorov-smirnov limit theorems for empirical distributions. Ann. Math. Stat. **19**(2), 177–189 (1948)
8. Justel, A., Pena, D., Zamar, R.: A multivariable kolmogorov-smirnov test of goodness of fit. Stat. Probab. Lett. **35**, 251–259 (1997)
9. Vidyasagar, M.: Learning and generalization: with applications to neural networks and control systems. Springer, London (2003)
10. Virginia Goss Tusher, R.T., Chu, G.: Significance analysis of microarrays applied to the ionizing radiation responses. Proc. Natl. Acad. Sci **98**(9), 5116–5121 (2001)
11. Efron, B., Tibshirani, R.: On testing the significance of a set of genes. Ann. Appl. Stat. **1**(1), 107–129 (2007)
12. Subramanian, A., et al.: Gene set enrichment analysis: a knowledge-based approach for interpreting genome-wide expression profiles. Proc. Natl. Acad. Sci. **102**(43), 15,545–15,550 (2005)
13. Tibshirani, R., et al.: Diagnosis of multiple cancer types by shrunken centroids of gene expression. Proc. Natl. Acad. Sci. **99**(10), 6567–6572 (2002)

Chapter 3
Inferring Gene Interaction Networks

Abstract This chapter contains the original research results on the monograph. We study the problem of reverse-engineering context-specific, genome-wide interaction networks from expression data. Two existing classes of methods, namely those based on mutual information and those based on Bayesian networks, are described first. Then a new algorithm, based on the so-called phi-mixing coefficient between random variables, is introduced. Unlike mutual information, the phi-mixing coefficient provides a directionally sensitive measure of the dependence between two random variables. The algorithm based on this new approach produces a gene interaction network in the form of a directed, strongly connected graph. The approach is validated on ChIP-seq data around the transcription factor ASCL1 in a lung cancer network.

Keywords Gene regulatory networks · Gene interaction networks · ARACNE · MINDy · CLR · Phixer · ASCL1 · NKX2-1 · PPARG · Lung cancer

3.1 Background

Each cell of a living organism contains a copy of its DNA. Thus in principle each cell contains all the genes of the organism, and each gene is in principle capable of producing all the gene products associated with it. It is therefore of vital interest to understand how precisely all the genes and gene products within a cell interact with each other. Such a network is referred to in the literature by various phrases, with 'gene regulatory network' (GRN) being among the most popular. For reasons discussed below, we prefer the phrase 'gene interaction network' (GIN). Moreover, though in principle the same genes are present in every cell, the manner in which each gene or gene product interacts with other genes or gene products, that is to say the GIN, can vary from one organ to another.[1] When we study diseased tissues, such as cancerous tumors, it will again be the case that the interactions within diseased cells will be different from those within normal cells in the same organ. In short, a plethora of interaction networks are needed to capture the variety of cells and their functioning.

By far the most popular representation of such an interaction pattern is a graph wherein the nodes are the genes and the edges denote the interactions between genes. The edges can either be directed or undirected, and they can either be weighted or unweighted. Ideally it would be beneficial to assign not just directionality to the interactions, but also weights, denoting the strength of the interaction. However, not all methods will produce graphs that are directed and/or weighted. There is one more issue. In biology, it can happen that the interaction between two genes is 'mediated' by a third gene. To cater to this possibility, sometimes one includes within the graphical representation both mediated edges as well as unmediated edges, as shown below.

Unmediated Interaction Mediated Interaction

Such graphical models are referred to in the literature by a variety of names, such as gene regulatory networks (GRNs), gene interaction networks (GINs), influence networks, correlation networks, co-expression networks, etc. In the present work, we prefer the name gene interaction network (GIN) over gene regulatory network (GRN), because in biology the phrase 'regulation' has a very precise meaning. If we say that gene A 'regulates' gene B, we are also expected to say *how* it does so, from a chemical and/or biological standpoint. However, as will be evident shortly, methods based on probabilistic analysis can at best infer the existence of some sort of interaction, its directionality, and in some cases, its weight. If the interaction of gene A on gene B is stronger than in the opposite direction, we can reasonably conclude that gene A has greater impact on the behavior of gene B than in the opposite direction. But this falls far short of what a biologist would mean by 'regulation'. For this reason, throughout we prefer to use the phrase gene interaction network (GIN). However, the phrase gene regulatory network (GRN) is equally, if not more, popular in the literature.

Figure 3.1 shows a small gene interaction network in mouse embryonic stem cells [1]. In this GIN, there are several nodes that have no incoming edges, and other nodes that have no outgoing edges. As a consequence, this graph is *not strongly connected*. Recall that a directed graph is said to be **strongly connected** if there exists a directed path between every pair of nodes. In reality, we would expect that any GIN would in fact be strongly connected, because it is highly implausible that there are some genes that are simply 'dangling'. On the other hand, if we think of

---

[1] Hereafter we shall avoid the unwieldy phrase 'genes or gene products' and shall instead say just 'genes'. However, proteins are also encompassed in the phrase 'gene products', and protein interaction networks (PINs) are therefore subsumed by the phrase GINs introduced a little later.

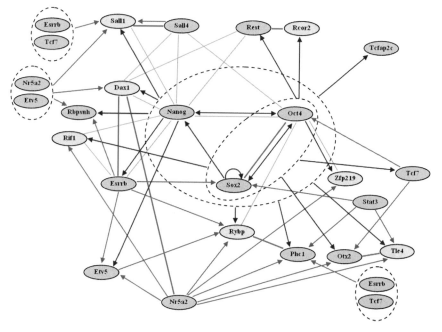

Fig. 3.1 A small GIN in mouse embryonic stem cells [1]

the above GIN, not as a complete entity, but as a subgraph of a much larger GIN, then it would be perfectly acceptable for the subgraph not to be strongly connected.

Usually small networks such as those in Fig. 3.1 are constructed by combining the outcomes of several painstaking experiments, each experiment (or set of experiments) designed to confirm the existence of one edge. As a consequence, one can accept with high confidence that all edges in a GIN constructed in this manner do exist in reality. However, human DNA contains about 22,000 genes. Consequently any 'genome-wide' GIN that includes all genes must of necessity contain millions of edges. It is simply infeasible to undertake enough experiments to infer the existence of so many edges, and one is forced to look for alternative approaches for reverse-engineering genome-wide GINs.

There are several existing databases of gene or protein interaction networks, either in the public domain or available from commercial vendors; see for example [2–6]. Invariably these networks are created by combining several individual networks such as the one in Fig. 3.1, reported in different publications in the literature. Indeed, it is common, especially for commercial vendors, to use text-mining or natural language processing (NLP) techniques to 'read' the scientific literature, thereby extract individual relationships between genes in the form of edges in the graph, and combine them into one or more large networks. The main shortcoming of this approach is that networks created in this fashion are *not context-specific*. Small networks of the type shown in Fig. 3.1, or individual edges reported in the literature, are derived under very specific experimental conditions. Combining these individual edges in a willy-nilly fashion while not keeping track of the underlying experimental conditions can lead to networks that are misleading and potentially erroneous.

Another shortcoming of existing databases of large networks is that, while they contain an impressive number of nodes and edges, they are still *not genome-wide*. The largest existing networks cover fewer than half of the known genes and/or proteins, and the number of interactions is at best in the 100,000s. The statistics on GINs cited in Fig. 2 of [7] justify this statement. This means that large parts of the genomic landscape are un- or under-explored.

Given that cancer is a very complex disease, it would be advantageous to be able to reverse-engineer, in an automated fashion, *context-specific, genome-wide GINs* on the basis of data that embraces all (or at least most) genes in the genome, with the data being derived under a common experimental setting. Methods for doing so are the topic of this chapter.

One can divide GIN models into two classes: static and dynamic. Dynamic GIN models usually consist of a system of ordinary differential equations (ODEs). See [8] and the references therein for exemplars of such an approach. Obviously, in order to generate such models, the experimental data must itself be temporally labeled. However, 'temporal' gene expression data is in reality a collection of ostensibly identical experiments terminated in a staggered fashion at different points of time. In the author's opinion, such data is often not reliable enough to permit the construction of accurate temporal models, unless the models are particularly simple. For this reason, the discussion in this work is focused on static GINs, where all quantities are in the steady-state.

The problem of inferring a GIN is that of reconstructing (or at least making a good model of) the GIN from experimental data, most commonly gene expression data. One of the main motivations for inferring GINs from data is very nicely spelled out in the perspective paper [9]:

> In the end, a good model of biological networks should be able to predict the behavior of the network under different conditions and perturbations and, ideally, even help us to engineer a desired response. For example, where in the molecular network of a tumor should we perturb with drug to reduce tumor proliferation or metastasis? Such a global understanding of networks can have transformative value, allowing biologists to dissect out the pathways that go awry in disease and then identify optimal therapeutic strategies for controlling them.

The paper [9] presents a set of three 'principles' and six 'strategies' for developing network models in cancer. The paper is well worth reading in its entirety. However, we note that Principle 1 is 'Molecular influences generate statistical relations in data', while Strategy 3 is 'Statistical identification of dysregulated genes and their regulators'. Given the scope of the present work, the discussion below is guided by these two observations.

A complete GIN is usually extremely complicated, with possibly tens of thousands of nodes, and millions of edges, often resembling a 'spider's web'. Figure 3.2 shows a part of the GIN corresponding to B lymphocytes, showing all the nearest neighbors of the proto-oncogene MYC, together with *some* (not all) of the neighbors of the neighbors of MYC; the figure corresponds to [10, Fig. 4] . We shall return to this example later.

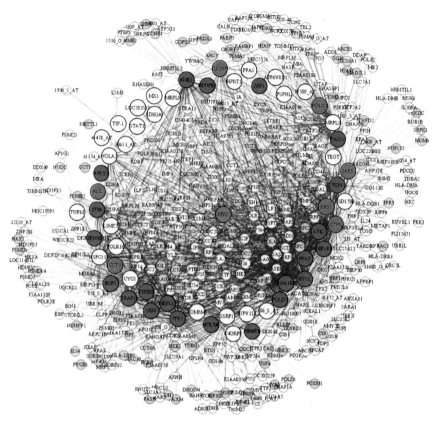

Fig. 3.2 The MYC Subnetwork [10, Fig. 4]

GINs have some very typical 'small world' features. For instance, simple arithmetic shows that with tens of thousands of nodes and millions of edges, the average connectivity of each node is in the double digit range. In reality however, the vast majority of nodes have connectivities in the single or low double-digit range, while a few nodes act as hubs and have connectivities in the high hundreds and possibly in the low thousands.

The problem at hand therefore is the reconstruction of a GIN on the basis of gene expression data, some (or most) of which could come from a public source such as the Gene Expression Omnibus (GEO) [11] or The Cancer Genome Atlas (TCGA) [12]. The main difference between GEO and TCGA is that the latter data sets are generated by large teams of researchers that work in concert, and are therefore more standardized than those in GEO, which is a repository for data sets associated with published papers. Even when the data has been painstakingly generated by personnel in some laboratory, the data is then immediately placed in GEO or another such publicly accessible source, so that the results can be verified by other research groups. Whether in TCGA or GEO, the data would contain expression levels of various gene products, obtained across multiple cell lines by various research teams (and all the lack of standardization that implies).[2] The data can be analyzed to study multiple genes or gene products in one cell line (lateral study), the same set of genes or gene products across multiple cell lines (longitudinal study), or both. In such studies, the number of gene products is often in the tens of thousands. However, the number of distinct cell lines rarely exceeds a few dozen, or a few hundred if one is extremely fortunate. Thus any statistical methodology must address this mismatch in dimension.

Another important aspect of the problem is that one rarely uses the 'raw' data coming out of experiments. As mentioned earlier, since biological experiments are not reproducible, every experiment includes some 'control' genes whose expression levels should be constant across experiments. Then the raw data from the various sets of experiments is normalized in such a way that the expression levels of the control genes is the same in all experiments. And then all the data is aggregated. Once this is done, the data for the remaining genes is 'smoothened' by centering, rescaling, linear to logarithmic transformation etc. The key point to note here is that each of these transformation is *one-to-one* and therefore invertible. Often the transformation is also *monotone*, in that it preserves the linear ordering of real numbers. The smoothened data then forms the input to the inference problem described next. It would therefore be highly desirable if the proposed solution to the inference problem is invariant under monotone transformations of the data. Indeed the new algorithm proposed here does have this property.

3.2 Problem Formulations

With that lengthy introduction, we are now ready to state formally the problems to be studied. In order to use statistical methods, we can think of the expression levels of the various genes or gene products as random variables $X_1, \ldots, X_n$, and the available data as consisting of samples $x_{ij}, i = 1, \ldots, n, j = 1, \ldots, m$. Ideally, we would like to compute the joint distribution function of the n random variables on the basis of the available data. However, the number of random variables n is far larger than the number of samples m and is likely to remain so for the foreseeable future. Thus, in any reasonable statistical sense, it is clearly impossible to infer the joint distribution of all n random variables, *unless one imposes some assumptions* on the nature of the joint distribution. The two specific techniques described below, namely the information-based approach and the Bayesian network approach, are distinguished by the assumptions they impose. It must be emphasized that the assumptions are imposed not so much because they are justified by biological realism, and more because they facilitate statistical analysis. In this sense, neither approach is fully

---

[2] Note that the data for a single cell line could itself be a compendium of data obtained through multiple experiments carried out at different times.

satisfactory. For this reason, after the review of existing methods, we present a new method that addresses some of the deficiencies of existing methods.

Irrespective of the assumptions made and the techniques used, given that the number of samples is much smaller than the number of random variables, the ultimate objective of statistical methods for inferring GINs must be restricted to unearthing dependences amongst various random variables. To put it another way, since the aim of finding a very precise formula for the joint distribution of the n random variables is unattainable because $n \gg m$, we must settle for determining whether one random variable X_i is influenced by another X_k. Let us now attempt to make precise this notion of 'being influenced'. At a very basic level, one could say that X_i is influenced by X_k if the two random variables X_i and X_k are not independent. But this is a very crude definition, so let us attempt to refine it. Suppose X_i is indeed influenced by X_k in the sense that X_i and X_k are not independent. The next level question one can ask is whether the influence is direct or indirect. In other words, is it the case that

$$\Pr\{X_i|X_j, j \neq i\} = \Pr\{X_i|X_j, j \neq i, j \neq k\}?$$

The above equation means that the conditional distribution of X_i given all other random variables $X_j, j \neq i$, is exactly the same as the conditional distribution of X_i given all random variables X_j other than X_k. For future use, we redefine notation and restate the above equation. Let $\mathcal{N}$ denote the finite set $\{1, \ldots, n\}$, and let us use the shorthand $\mathcal{N} \setminus i$ instead of the more precise $\mathcal{N} \setminus \{i\}$. Then the question to which we seek a yes or no answer can be stated as

$$\Pr\{X_i|X_{\mathcal{N}\setminus i}\} = \Pr\{X_i|X_{\mathcal{N}\setminus i,k}\}? \tag{3.1}$$

If the above equation holds, then it means that, while X_k does indeed influence X_i, the influence is indirect.

The question can be made even more explicit by using the notion of conditional independence, defined next.

Definition 3.1. Suppose X, Y, Z are random variables assuming values in finite sets $\mathbb{A}, \mathbb{B}, \mathbb{C}$ respectively. Then X and Z are **conditionally independent** given Y, denoted by $(X \perp Z)|Y$, if the following property holds: For all $i \in \mathbb{A}, j \in \mathbb{B}, k \in \mathbb{C}$, it is the case that

$$\Pr\{X = i \& Z = k|Y = j\} = \Pr\{X = i|Y = j\} \cdot \Pr\{Z = k|Y = j\}. \tag{3.2}$$

If (3.2) holds, then it is easy to verify that, for all $S \subseteq \mathbb{A}, j \in \mathbb{B}, U \subseteq \mathbb{C}$, it is the case that

$$\Pr\{X \in S \& Z \in U|Y = j\} = \Pr\{X \in S|Y = j\} \cdot \Pr\{Z \in U|Y = j\}. \tag{3.3}$$

However, in general (3.2) *does not imply* that, for all $S \subseteq \mathbb{A}, T \subseteq \mathbb{B}, U \subseteq \mathbb{C}$,

$$\Pr\{X \in S \& Z \in U | Y \in T\} = \Pr\{X \in S | \in Y \in T\} \cdot \Pr\{Z \in U | Y \in T\}. \qquad (3.4)$$

If (3.4) is true, then by choosing $T = \mathbb{B}$, we can conclude from (3.2) that

$$\Pr\{X \in S \& Z \in U\} = \Pr\{X \in S\} \cdot \Pr\{Z \in U\},$$

i.e. that X and Z are independent random variables. Conversely, if X and Z are independent, then clearly (3.4) is true. Note that conditional independence is a weaker property than independence.

Our immediate objective is to recast the desired relationship (3.1) in terms of conditional independence. The next theorem provides a way of doing so.

Theorem 3.1. X, Z *are conditionally independent given* Y *if and only if, for all* $i \in \mathbb{A}, j \in \mathbb{B}, k \in \mathbb{C}$, *it is the case that*

$$\Pr\{X = i | Z = k \& Y = j\} = \Pr\{X = i | Y = j\}. \qquad (3.5)$$

Proof. In the interests of simplicity, we will write $\Pr\{X|ZY\}$ for $\Pr\{X = i | Z = k \& Y = j\}$, and so on. Then by the definition of conditional probability, we have that

$$\Pr\{XZY\} = \Pr\{X|ZY\} \Pr\{Z|Y\} \Pr\{Y\}. \qquad (3.6)$$

However, if $(X \perp Z) | Y$, then it follows from (3.2) that

$$\begin{aligned}
\Pr\{XZY\} &= \Pr\{XZ|Y\} \Pr\{Y\} \\
&= \Pr\{X|Y\} \Pr\{Z|Y\} \Pr\{Y\}. \qquad (3.7)
\end{aligned}$$

Comparing the right sides of (3.6) and (3.7) shows that (3.5) must be true. The proof in the other direction consists simply of reversing the above argument. $\qquad \square$

In view of Theorem 3.1, it is evident that (3.1) can be restated in terms of conditional independence. Specifically

$$\Pr\{X_i | X_j, j \neq i\} = \Pr\{X_i | X_j, j \neq i, j \neq k\} \iff (X_i \perp X_k) | X_{\mathcal{N} \setminus i,k}. \qquad (3.8)$$

The above equation can be interpreted in terms of the GIN associated with the variables $X_i, i \in \mathcal{N}$. If there is no edge of the form (i, k) or (k, i) in the GIN, then the removal of all nodes other than i and k would cause the graph to be disconnected; therefore (3.8) holds. Taking the contrapositive shows that if (3.8) does not hold, then the GIN must contain at least one of the two edges of the form (i, k) or (k, i). However, verifying (3.8) for all $X_{\mathcal{N} \setminus i,k}$ is impractical. So we settle for a simplified version of this condition. Given the data set, for each triplet $(i, j, k) \in \mathcal{N}^3$ of pairwise distinct indices, let us determine whether or not it is true that $(X_i \perp X_k) | X_j$. If indeed it is the case that $(X_i \perp X_k) | X_j$, then every path from i to k and vice versa must pass through j. In particular, there is no edge of the form (i, k) or (k, i) in the GIN. Taking the contrapositive shows that if $(X_i \not\perp X_k)|X_j$, then there is at least one path from i to k or vice versa that does not pass through j. Using the Occam's razor principle, namely using the simplest possible explanation, we would conclude that GIN must contain at least one of the two edges of the form (i, k) or (k, i). Notice however that the last conclusion is not a mathematical certainty, but rather a plausibility argument based on the Occam's razor principle.

3.3 Methods Based on Mutual Information

One way to approach the issue of whether X_j influences X_i is to compute their mutual information. Let us switch notation and suppose that X, Y are random variables assuming values in finite sets $\mathbb{A}, \mathbb{B}$ respectively. Let μ, ν, θ denote the distribution of X, the distribution of Y, and the joint distribution of X and Y, respectively. Then the quantity

$$H(X) = H(\mu) = -\sum_{i \in \mathbb{A}} \mu_i \log \mu_i$$

is called the **Shannon entropy** of X or μ. Note that we make no distinction between the entropy of a probability distribution μ and the entropy of a random variable X having the probability distribution μ. Next,

$$I(X, Y) = H(X) + H(Y) - H(X, Y)$$

is called the **mutual information** between X and Y. An equivalent formula is

$$I(X, Y) = \sum_{i \in \mathbb{A}} \sum_{j \in \mathbb{B}} \theta_{ij} \log \frac{\theta_{ij}}{\mu_i \nu_j}.$$

Note that mutual information is symmetric: $I(X, Y) = I(Y, X)$. Also, $I(X, Y) = 0$ if and only if X, Y are independent random variables. Finally, if $f : \mathbb{A} \to \mathbb{A}', g : \mathbb{B} \to \mathbb{B}'$ are one-to-one and onto maps then $I(f(X), g(Y)) = I(X, Y)$. Thus in particular, monotone maps of random variables leave the entropy and mutual information invariant.

One of the first attempts to use mutual information to construct GINs is in [13], which introduces 'influence networks.' In this approach, given m samples each for n random variables X_1 through X_n, one first computes the pairwise mutual information $I(X_i, X_j)$ for all $i, j, j \neq i$, that is, $n(n-1)/2$ pairwise mutual informations. Then X_i and X_j are said to influence each other if the computed $I(X_i, X_j)$ exceeds a certain threshold. Note that, since mutual information is symmetric, in case $I(X_i, X_j)$ does exceed the threshold, all one can say is that X_j influences X_i, or vice versa, or perhaps both. In other words, it is not possible to infer any 'directionality' to the influence if one uses mutual information (or for that matter any other symmetric quantity) to infer dependence. Another detail is note is that in fact one cannot compute the 'true' mutual information because one does not know the true joint distributions of X_i, X_j. Instead, one has to compute an 'empirical' approximation to $I(X_i, X_j)$ on the basis of the samples. In [13], this is done by grouping the observed expression levels into ten histograms, thus effectively quantizing each random variable into one of ten values. Presumably the number of bins was chosen to be ten because in [13] because they had 79 samples. In cases where the number of samples is smaller, one would obviously have to use fewer bins.

The major drawback of the influence networks approach proposed in [13] is that it is not able to discriminate between direct and indirect influence. As a result, the influence network constructed using mutual information is overly dense, because it *fails* to have an edge between nodes i and j if and only if X_i and X_j are independent (or if one uses empirically computed estimates for the mutual information and a threshold, nearly independent). To get more meaningful results, it is necessary to *prune* this first-cut influence network by deleting an edge between nodes i and j if the influence is indirect, that is, if (3.1) holds.

To achieve this objective, an algorithm called ARACNE is proposed in [14]. The basis of this algorithm is the assumption that the joint probability distribution of all n variables factors into a product of terms involving at most two variables at a time. This special feature makes it possible to invoke a bound known as the data processing inequality to prune the first-cut influence network.

Now we describe the ARACNE algorithm.[3] To make the ideas clear, let us suppose that the random variable X_i assumes values in a finite alphabet $\mathbb{A}_i$, which can depend on i. Define $\mathbb{A} = \prod_{i=1}^{n} \mathbb{A}_i$, and let $\mathbf{x}$ denote the n-tuple $(x_1, \ldots, x_n) \in \mathbb{A}$. Similarly let $\mathbf{X}$ denote $(X_1, \ldots, X_n)$. Then the joint distribution of all n random variables is the function $\phi : \mathbb{A} \to [0, 1]$ defined by

$$\phi(\mathbf{x}) = \Pr\{\mathbf{X} = \mathbf{x}\}. \tag{3.9}$$

Now let $\mathcal{N} = \{1, \ldots, n\}$, and let

$$\mathcal{D} = \{(i, j) : 1 \leq i < j \leq n\}.$$

Then the assumption that underlies the ARACNE algorithm is that the function ϕ has the form

$$\phi(\mathbf{x}) = \frac{1}{Z} \prod_{i \in \mathcal{N}} \psi_i(x_i) \cdot \prod_{(i,j) \in \mathcal{D}} \phi_{ij}(x_i, x_j), \tag{3.10}$$

where

$$Z = \sum_{\mathbf{x} \in \mathbb{A}} \left[\prod_{i \in \mathcal{N}} \psi_i(x_i) \cdot \prod_{(i,j) \in \mathcal{D}} \phi_{ij}(x_i, x_j) \right]$$

---

[3] Note that language used here is not identical to that in [14] but is mathematically equivalent.

is a normalizing constant. Note that in the statistical mechanics terminology employed in [14], the quantity $\log \phi(\cdot)$ is called the 'Hamitonian,' and the assumption is that the Hamiltonian is the *sum* of terms involving only individual x_i, or pairs (x_i, x_j), but no higher order terms.

Suppose we associate an undirected graph with the distribution in (3.10) by inserting an edge[4] between nodes i and j if the function ϕ_{ij} is not identically zero. In the worst case, if every such function is not identically zero, we would wind up with a complete graph with n nodes, where every node is connected to every other node. This is clearly not desirable. So the authors of [14] set out to find a simpler representation of the data than a complete graph. In doing so, they build upon the work of [15], where the objective is to find the best possible approximation to a given probability distribution $\phi(\cdot)$ (not necessarily of the form (3.10)) in terms of a distribution of the form (3.10) where $\phi_{ij} \not\equiv 0$ for exactly $n-1$ pairs. The criterion used to define 'best possible' is the relative entropy or the Kullback-Leibler divergence [16, p. 19]. Specifically, if ϕ is the original distribution and θ is its approximation, then the quantity to be minimized is

$$H(\phi\|\theta) = \sum_{\mathbf{x}} \phi(\mathbf{x}) \log \frac{\phi(\mathbf{x})}{\theta(\mathbf{x})}.$$

This problem has a very elegant solution, as shown in [15]. Starting with the given distribution ϕ, first compute all $n(n-1)/2$ pairwise mutual informations $I(X_i, X_j), j \neq i$. Then sort them in decreasing order. Suppose $I(X_{i_1}, X_{i_2})$ is the largest; then place an edge between nodes i_1 and i_2. Suppose $I(X_{i_3}, X_{i_4})$ is the next largest. Then create an edge between nodes i_3 and i_4. In general, at step k, suppose $I(X_{i_{2k-1}}, X_{i_{2k}})$ is the k-th largest mutual information. Then create an edge between nodes i_{2k-1} and i_{2k}, unless doing so would create a loop; in the latter case, go on to the next largest mutual information. Do this precisely $n-1$ times. The result is an undirected graph with n nodes, $n-1$ edges, and no cycles—in other words, a tree.

The authors of [14] build upon this approach by invoking the following result. If X_i, X_k are conditionally independent given X_j, then the so-called 'data processing inequality' [16, p. 35] states that

$$I(X_i, X_k) \leq \min\{I(X_i, X_j), I(X_j, X_k)\}. \tag{3.11}$$

Accordingly, the ARACNE algorithm initially constructs an influence network as in [13]. Then for each triplet (i, j, k) of pairwise distinct indices, the three quantities $I(X_i, X_j), I(X_i, X_k), I(X_j, X_k)$ are compared; the smallest among the three is deemed to arise from an indirect interaction, and the corresponding edge is deleted.

From the above description, it is easy to deduce the following fact: A network produced by the ARACNE algorithm will never contain a complete subgraph with three nodes. In other words, if there exist edges between nodes i and j, and between nodes j and k, then there will never be an edge between nodes i and k. From the

---

[4] Note that since the graph is undirected, it is not necessary to specify the direction.

standpoint of biology, this means that if gene i influences (or is influenced by) two other genes j and k, then perforce genes j and k must be conditionally independent given the activity level of gene i.

Note that the network that results from applying the ARACNE algorithm does not depend on where we start the pruning. To illustrate, consider a very simple-minded network with four nodes as shown below.

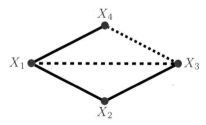

Suppose that

$$I(X_1, X_3) \leq \min\{I(X_1, X_2), I(X_2, X_3)\}.$$

In accordance with the algorithm, the link from X_1 to X_3 is discarded (and thus shown as a dashed line). Now suppose in addition that

$$I(X_3, X_4) \leq \min\{I(X_1, X_3), I(X_1, X_4)\}.$$

Then the edge from X_3 to X_4 is also deleted. It is easy to verify that if we had examined the triplets in the opposite order we would still end with the same final graph.

The ARACNE algorithm has been applied to the problem of reverse-engineering regulatory networks of human B cells in [10]. A total of 336 gene expression profiles for 9,563 genes were used. Only about 6,000 genes had sufficient variation in expression levels to permit the computation of mutual information. To illustrate the network that results from applying the algorithm, the authors depict how it looks in the vicinity of the proto-oncogene MYC.[5] The ARACNE algorithm showed that MYC had 56 nearest neighbors, and these 56 neighbors had 2,007 other genes that were not neighbors of MYC. Thus at a distance of two steps, MYC contained more than 2,000 of the roughly 6,000 genes in the network. The overall network had about 129,000 interactions (edges), or about 20 per node on average. However, just 5 % of the 6,000 nodes accounted for 50,000 edges, or about 40 % of the total, thus demonstrating the 'small world' nature of the GIN that results from the algorithm. Figure 3.2 shows the 56 neighbors and another 444 most significant second neighbors of MYC.

Thus far the methods described generate GINs with only unmediated edges. To construct GINs with mediated edges, one follows the same approach as in ARACNE, except that instead of using the mutual information $I(X_i, X_j)$, one uses the conditional

---

[5] Medterms [17] defines a proto-oncogene as "A normal gene which, when altered by mutation, becomes an oncogene that can contribute to cancer," and an oncogene as "A gene that played a normal role in the cell as a proto-oncogene and that has been altered by mutation and now may contribute to the growth of a tumor."

mutual information $I(X_i|X_l, X_j|X_l)$. Since the conditional mutual information also satisfies a data processing inequality of the form (3.11), the same reasoning can be applied to prune an initially overly dense network. This algorithm, based on conditional mutual information, is referred to as MINDy and is proposed in [18]. An essentially similar algorithm is proposed in [19].

In either ARACNE or MINDy, it is obvious that the most time-consuming step is the computation of all pairwise mutual informations. In [14], the authors take the given samples, and then fit them with a two-dimensional Gausian kernel for each pair of random variables. Then a copula transform is applied so that the sample space is the unit square, and the marginal probability distribution of each random variable is the uniform distribution.[6] In [22], a window-based approach is presented for computing pairwise mutual information that is claimed to result in roughly an order of magnitude reduction in the computational effort. For instance, for the B lymphocyte network studied in [10], the original ARACNE computation is claimed to take 142 h of computation, while the method proposed in [22] is claimed to take only 23 h.[7] Finally, in a very recent paper [23], the authors bin the samples into just three bins irrespective of how many samples there are, and propose a highly efficient parallel architecture for computing the pairwise mutual informations. While the proposed architecture is very innovative, it appears to the present author that quantizing the expression values into just three bins could result in misleading conclusions, because the gene expression level is essentially a *real-valued* random variable. Just to look ahead, in Sect. 3.6 we propose a new algorithm based on computing the so-called ϕ-mixing coefficient between two random variables. Unlike the mutual information, we present a *closed-form formula* for the ϕ-mixing coefficient between two random variables, so that its implementation is extremely efficient, even on a garden variety desktop workstation.

3.4 Methods Based on Bayesian Networks

In this section we discuss the Bayesian network-based approach to inferring GINs. Bayesian networks have been used in artificial intelligence for many decades, and [24] is the classic reference for that particular application domain. The Bayesian approach to inferring GINs appears to have been pioneered in [25]. This was followed up by other work [26] and a survey is given in [27].

As before, the problem is to infer the joint distribution of n random variables $X_1, \ldots, X_n$, based on m independent samples of each random variable. For any set of random variables, it is possible to write their joint distribution as a product of conditional distributions. For two variables X_1, X_2, we can write

$$\Pr\{X_1, X_2\} = \Pr\{X_1\} \cdot \Pr\{X_2|X_1\},$$

---

[6] The notion of a copula was introduced in [20]. See [21] for an excellent introduction to the topic.

[7] It is interesting to note that in a preprint version of [22], their method is claimed to take only 1.6 h.

and we can also write

$$\Pr\{X_1, X_2\} = \Pr\{X_2\} \cdot \Pr\{X_1|X_2\}.$$

If there are n random variables $X_1, \ldots, X_n$, then we can write

$$\Pr\{X_{\mathcal{N}}\} = \prod_{i=1}^{n} \Pr\{X_i|X_j, 1 \leq j \leq i-1\},$$

where $X_{\mathcal{N}}$ denotes $(X_1, \ldots, X_n)$. More generally, let π be any permutation on $\{1, \ldots, n\}$. Then we can also write

$$\Pr\{X_{\mathcal{N}}\} = \prod_{i=1}^{n} \Pr\{X_{\pi(i)}|X_{\pi(1), \ldots, \pi(i-1)}\}. \tag{3.12}$$

Since the above expression is valid for *every* permutation π, we should choose to order the variables in such a way that the various conditional probabilities become as simple as possible. In essence, this is the basic idea behind Bayesian networks.

Suppose now that $\mathcal{G}$ is an acyclic directed graph with n vertices. Note the total contrast with the assumptions in methods based on mutual information. In that setting, $\mathcal{G}$ is an undirected graph, so that edges can be thought of as being bidirectional. In the present setting, not only are edges unidirectional, but no cycles are permitted. In other words, both the situations shown below are ruled out in the Bayesian network paradigm.

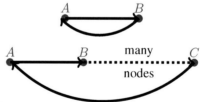

Let us think of an edge (i, j) as being *from* node i *to* node j, and let $\mathcal{E}$ denote the set of edges in $\mathcal{G}$. Since the graph is assumed to be acyclic, with each node i we can unambiguously associate its ancestor set $A(i)$ and its successor set $S(i)$ defined by

$$A(i) = \{k : (k, i) \in \mathcal{E}\}, S(i) = \{j : (i, j) \in \mathcal{E}\}. \tag{3.13}$$

Since the graph is both directed as well as acyclic, it is obvious that $A(i)$ and/or $S(i)$ may be empty for some indices i. In some circles, a node i is referred to as a 'source' if $A(i)$ is empty, and as a 'sink' if $S(i)$ is empty. Recall the notation $X_i \perp X_j$ if X_i and X_j are independent, and the notation $(X_i \perp X_j)|X_k$ if X_i and X_j are conditionally independent given X_k.

Definition 3.2. A set of random variables $X_1, \ldots, X_n$ is said to be a **Bayesian network** with respect to a directed acyclic graph $\mathcal{G}$ if

$$(X_i \perp X_j)|X_{A(i)}, \quad \forall j \notin S(i). \tag{3.14}$$

In words, a set of random variables $X_1, \ldots, X_n$ is a Bayesian network with respect to $\mathcal{G}$ if, for a fixed index i, the associated r.v. X_i is conditionally independent of X_j for all nonsuccessors j, given the values of X_k for all ancestors k of i. It is easy to see that if the set of random variables $X_1, \ldots, X_n$ forms a Bayesian network with respect to the directed acyclic graph $\mathcal{G}$, then the joint probability distribution factors as follows:

$$\Pr\{X_{\mathcal{N}}\} = \prod_{i=1}^{n} \Pr\{X_i|X_{A(i)}\}, \tag{3.15}$$

where the conditional probability of X_i is taken to be the unconditional probability if the set $A(i)$ is empty. Compare (3.15) with (3.12).

The formula (3.15) demonstrates one of the main attractions of the Bayesian network model. For each source node i, the unconditional probability of X_i can be computed directly from the data. (It is obvious that if i, j are both source nodes, then $X_i \perp X_j$.) Then, using (3.15), the conditional probability computation of any intermediate X_i can be propagated along the graph. This is the feature that makes Bayesian networks so popular in AI circles. In an AI application, the underlying graph $\mathcal{G}$ is posited *a priori* on the basis of expert opinion or prior knowledge. In the context of inferring GRNs, the graph itself is unknown, and the objective of the exercise is to determine 'the best possible fit' to the observed data. The paper [25] contains a detailed discussion of this problem, as does the tutorial [28]; So for present purposes we content ourselves with just a quick overview.

The problem of modeling a set of expression data using a Bayesian network can be divided into two questions. First, what is the graph $\mathcal{G}$ that is used to model the data (i.e., the dependence structure among the random variables)? Second, once the graph $\mathcal{G}$ has been chosen, how can one find the best possible fit to the expression data by a suitable choice of the various conditional probabilities in (3.15)? In answering the second question, one again needs to make a distinction between parametric models, where the various conditional probabilities are specified as known functions of an unknown parameter $\theta \in \Theta$ where Θ is specified ahead of time, and nonparametric models in which case no such form is assumed. Strictly speaking, the classical Bayesian paradigm applies to the use of parametric models with the dependence structure specified beforehand. In such a case, it is assumed that the parameter θ has a known prior distribution, and that the data set, call it D, is generated using some unknown probability distribution. Then the parameter θ is chosen so as to maximize the posterior probability $\Pr\{\theta|D\}$. The Bayesian approach consists of observing that

$$\Pr\{\theta|D\} = \frac{\Pr\{D|\theta\} \cdot \Pr\{\theta\}}{\Pr\{D\}}.$$

Hence

$$\log \Pr\{\theta|D\} = \log Pr\{D|\theta\} + \log \Pr\{\theta\} - \log \Pr\{D\}.$$

In the above equation, $\Pr\{D\}$ can be treated as a constant, since it does not depend on θ. In principle, the same approach can also be extended to answer the first question as well, namely the choice of the directed graph $\mathcal{G}$ that is used to model the data. However, since the number of possible directed acyclic graphs in n nodes increases far too quickly with n, this approach may not be feasible, unless one restricts attention to a very small subset of all possible directed acyclic graphs on n nodes.

3.5 A Unified Interpretation

The two approaches described above can be put into some sort of common framework. Suppose $X_1, \ldots, X_n$ are random variables assuming values in finite sets $\mathbb{A}_1, \ldots, \mathbb{A}_n$ respectively. Let $X_{\mathcal{N}}$ denote $(X_1, \ldots, X_n)$, and let $\mathbb{A}$ denote $\prod_{i=1}^{n} \mathbb{A}_i$. Finally, let $\mathbf{x} \in \mathbb{A}$ denote a value that $X_{\mathcal{N}}$ can assume, and let, as before,

$$\phi(x) = \Pr\{X_{\mathcal{N}} = \mathbf{x}\}$$

denote the joint probability distribution. Then one can ask two specific questions: First, if $\phi(\mathbf{x})$ has certain product form, does this imply any kind of dependence structure on the random variables? Second, and conversely, if the random variables have some kind of dependence structure, does this imply that the joint distribution has a specific form? It turns out that the first question is very easy to answer, while the second one is more difficult.

Accordingly, suppose first that $\mathcal{G}$ is a graph with n nodes. For the moment we neither assume that the graph is symmetric nor that it is acyclic. It is a directed graph (unlike in ARACNE) and may contain cycles (unlike in the case of Bayesian networks). Let $\mathcal{N}$ denote $\{1, \ldots, n\}$, the set of nodes in the graph, and suppose $C_1, \ldots, C_k$ are subsets of $\mathcal{N}$ that together cover $\mathcal{N}$. In other words,

$$\bigcup_{l=1}^{k} C_l = \mathcal{N}.$$

Besides the covering property, no other assumptions are made about the nature of the C_l. For each C_l, define

$$X_{C_l} = (X_j, j \in C_l), \quad \mathbb{A}_{C_l} = \prod_{j \in C_l} \mathbb{A}_j.$$

The possible value $\mathbf{x}_{C_l} \in \mathbb{A}_{C_l}$ is defined analogously. Next, define

$$D(i) = \bigcup \{C_l : i \in C_l\}, \quad T(i) = D(i) \setminus \{i\}.$$

Thus $D(i)$ consists of the union of all C_l that contain i. Note that, due to the covering property of the sets C_l, there is at least one C_l that contains i, whence $D(i)$ is nonempty and contains i. Thus $T(i)$ is well-defined, though it could be empty. With these definitions, the following result ensues.

Theorem 3.2. *Suppose there exist functions* $\phi_l, l = 1, \ldots, k$ *such that*

$$\phi(\mathbf{x}) = \frac{1}{Z} \prod_{l=1}^{k} \phi_l(\mathbf{x}_{C_l}), \tag{3.16}$$

where

$$Z = \sum_{\mathbf{x} \in \mathbb{A}} \prod_{l=1}^{k} \phi_l(\mathbf{x}_{C_l})$$

is a normalizing constant. Then

$$\Pr\{X_i | X_{\mathcal{N} \setminus i}\} = \Pr\{X_i | X_{T(i)}\}. \tag{3.17}$$

An equivalent way of stating the theorem, which makes it resemble the definition of a Bayesian network is this: Suppose the joint distribution $\phi(\mathbf{x})$ can be factored as in Theorem 3.2. Then

$$(X_i \perp X_j) | X_{T(i)} \ \forall j \notin \mathcal{N} \setminus D(i). \tag{3.18}$$

Proof.   From the definition of conditional probability, it follows that

$$\Pr\{X_i = x_i | X_{\mathcal{N} \setminus i} = \mathbf{x}_{\mathcal{N} \setminus i}\} = \frac{\Pr\{X_{\mathcal{N}} = \mathbf{x}\}}{\Pr\{X_{\mathcal{N} \setminus i} = \mathbf{x}_{\mathcal{N} \setminus i}\}}.$$

Substituting from (3.16) leads to

$$\Pr\{X_{\mathcal{N}} = \mathbf{x}\} = \frac{1}{Z} \prod_{l=1}^{k} \phi_l(\mathbf{x}_{C_l}) = \prod_{i \notin C_l} \phi_l(\mathbf{x}_{C_l}) \frac{1}{Z} \prod_{i \in C_l} \phi_l(\mathbf{x}_{C_l}).$$

Similarly

$$\Pr\{X_{\mathcal{N} \setminus i} = \mathbf{x}_{\mathcal{N} \setminus i}\} = \frac{1}{Z} \sum_{x_i \in \mathbb{A}_i} \prod_{l=1}^{k} \phi_l(\mathbf{x}_{C_l}) = \prod_{i \notin C_l} \phi_l(\mathbf{x}_{C_l}) \sum_{x_i \in \mathbb{A}_i} \frac{1}{Z} \prod_{i \in C_l} \phi_l(\mathbf{x}_{C_l}).$$

Note that the term $\prod_{i \notin C_l} \phi_l(\mathbf{x}_{C_l})$ is common to both expressions. So

$$\Pr\{X_i = x_i | X_{\mathcal{N}\backslash i} = \mathbf{x}_{\mathcal{N}\backslash i}\} = \frac{\Pr\{X_{\mathcal{N}} = \mathbf{x}\}}{\Pr\{X_{\mathcal{N}\backslash i} = \mathbf{x}_{\mathcal{N}\backslash i}\}}$$

$$= \frac{(1/Z) \prod_{i \in C_l} \phi_l(\mathbf{x}_{C_l})}{(1/Z) \sum_{x_i \in \mathbb{A}_i} \prod_{i \in C_l} \phi_l(\mathbf{x}_{C_l})}.$$

Note that $X_{\mathcal{N}\backslash D(i)}$ does not appear in either the numerator or the denominator. Hence we can sum over all $\mathbf{x}_{\mathcal{N}\backslash D(i)}$ and the ratio would be unchanged. In other words,

$$\Pr\{X_i = x_i | X_{\mathcal{N}\backslash i}\} = \frac{(1/Z) \sum_{\mathbf{x}_{\mathcal{N}\backslash D(i)}} \prod_{i \notin C_l} \phi_l(\mathbf{x}_{C_l}) \prod_{i \in C_l} \phi_l(\mathbf{x}_{C_l})}{(1/Z) \sum_{\mathbf{x}_{\mathcal{N}\backslash D(i)}} \sum_{x_i \in \mathbb{A}_i} \prod_{i \notin C_l} \phi_l(\mathbf{x}_{C_l}) \prod_{i \in C_l} \phi_l(\mathbf{x}_{C_l})}$$

$$= \frac{\Pr\{X_{D(i)} = \mathbf{x}_{D(i)}\}}{\Pr\{X_{T(i)} = \mathbf{x}_{T(i)}\}}$$

$$= \Pr\{X_i = x_i | X_{T(i)} = \mathbf{x}_{T(i)}\}.$$

This is the desired conclusion.   □

With suitable conventions, both the Bayesian network and the undirected graph can be put into the above dependence structure. However, the converse of the above theorem is false in general. Even if (3.17) holds, it does not readily follow that the joint distribution factors in the form (3.16). To obtain a proper converse, we introduce the notion of a Markov random field and present the Hammersley-Clifford theorem. Suppose as before that $X_1, \ldots, X_n$ are random variables assuming values in their respective finite alphabets (which need not be the same). Next, in contrast with Theorem 3.2, assume that $\mathcal{G}$ is an *undirected* graph with n nodes. For each node i, let $N(i)$ denote the set of neighbors of i; thus $N(i)$ consists of all nodes j such that there is an edge between nodes i and j.

Definition 3.3. A set of random variables $X_1, \ldots, X_n$ is said to be a **Markov random field** with respect to a graph $\mathcal{G}$ with n nodes if

$$\Pr\{X_i | X_{\mathcal{N}\backslash i}\} = \Pr\{X_i | X_{N(i)}\}, \ \forall i. \tag{3.19}$$

In words, a set of random variables $X_1, \ldots, X_n$ is a Markov random field with respect to $\mathcal{G}$ if and only if the conditional distribution of each random variable X_i depends only on its neighbors X_k, $k \in N(i)$. Note that (3.17) reduces to (3.19) if we define $T(i) = N(i)$.

A closely related notion is that of a Gibbs distribution. To define this notion, let us recall that a **clique** of an undirected graph is a maximal completely connected subgraph.

Definition 3.4. Suppose $X_1, \ldots, X_n$ are random variables and that $\mathcal{G}$ is an undirected graph with n nodes. Let $C_1, \ldots, C_k$ denote the cliques of $\mathcal{G}$. Then the random variables $X_1, \ldots, X_n$ are said to have a **Gibbs distribution** with respect to the graph $\mathcal{G}$ if their joint distribution ϕ satisfies

Fig. 3.3 Illustration of the
Hammersley-Clifford theorem

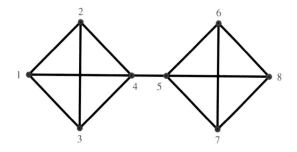

$$\phi(\mathbf{x}) = \prod_{l=1}^{k} \phi_l(\mathbf{x}_{C_l}). \tag{3.20}$$

In words, the random variables $X_1, \ldots, X_n$ have a Gibbs distribution with respect to $\mathcal{G}$ if the joint distribution of all n variables factors as a product of simpler joint distributions, one for each clique of $\mathcal{G}$.

A fundamental result known as the Hammersley-Clifford theorem connects the two concepts (Fig. 3.3).

Theorem 3.3. *Suppose the joint distribution $\phi(\mathbf{x})$ of a set of random variables is strictly positive for all $\mathbf{x}$. Then they form a Markov random field if and only if the joint distribution is a Gibbs distribution.*

Though this theorem is credited to Hammersley and Clifford, their original manuscript is somewhat inaccessible. A proof of this theorem can be found in [29] as well as several textbooks. Note that the proof in one direction is easy: If the joint distribution is Gibbs, then the random variables form a Markov random field, and one does not require the assumption that the joint distribution is positive in order to prove this. Indeed, this is captured in Theorem 3.2. Therefore the real import of the theorem is in the opposite direction. To prove it in this direction, the strict positivity of the joint distribution is an essential requirement.

Example 3.1. As an illustration of the application of the Hammersley-Clifford theorem, consider the network shown in Fig. 3.1. In this network there are three cliques, namely

$$C_1 = \{1, 2, 3, 4\}, C_2 = \{4, 5\}, C_3 = \{5, 6, 7, 8\}.$$

Accordingly, a set of random variables $\{X_1, \ldots, X_8\}$ is a Markov random field with respect to this graph if and only if the joint distribution of all eight variables factors into a product of the form

$$\phi(x_1, \ldots, x_8) = \prod \phi_1(x_1, x_2, x_3, x_4)\phi_2(x_4, x_5)\phi_3(x_5, x_6, x_7, x_8).$$

3.6 A New Algorithm Based on the Phi-Mixing Coefficient

The ARACNE algorithm discussed in Sect. 3.3 begins with a complete but undirected graph, and then prunes it via two steps. First, any pair of nodes for which the corresponding mutual information is smaller than a certain threshold is eliminated; this is the approach of [13]. Then further pruning of the remaining edges takes place using the data processing inequality (3.11). This suggests that, in order to infer gene interaction networks where the edges are both directed and weighted, the broad contours of the ARACNE algorithm can be retained, but the mutual information should be replaced by another quantity that both provides a directed measure of dependence between random variables and also satisfies an analog of the data processing inequality. In this section, it is shown that such a measure is provided by the so-called ϕ-mixing coefficient between random variables. The section is organized as follows. First, the ϕ-mixing coefficient is defined, and the details of computing it are discussed. Then it is shown that the ϕ-mixing coefficient satisfies an analog of the data processing inequality. With these two steps in place, the new algorithm is described. Finally, a case study in reverse-engineering a lung cancer GIN is presented.

3.6.1 The Phi-Mixing Coefficient: Definition and Computation

The notion of mixing originated in an attempt to establish the law of large numbers for stationary stochastic processes that are not i.i.d. Three notions of mixing are popularly used, namely α-mixing, β-mixing, and ϕ-mixing. General definitions of the α-, β- and ϕ-mixing coefficients of a stationary stochastic process can be found, among other places, in [30, pp. 34–35]. In reality, it is possible to define mixing coefficients between a pair of random variables, instead of for a stochastic process. This is the approach adopted in [31], which is an excellent (though terse) reference for mixing properties and inequalities. In the present work, we shall be interested only in the ϕ-mixing coefficient. This notion was introduced by Ibragimov [32].

Definition 3.5. Suppose X and Y are random variables assuming values in finite sets $\mathbb{A} = \{1, \ldots, n\}$ and $\mathbb{B} = \{1, \ldots, m\}$ respectively. Then the ϕ-mixing coefficient $\phi(X|Y)$ is defined as

$$\phi(X|Y) := \max_{S \subseteq \mathbb{A}, Y \subseteq \mathbb{B}} |\Pr\{X \in S|Y \in T\} - \Pr\{X \in S\}|. \tag{3.21}$$

Thus $\phi(X|Y)$ is the maximum difference between the conditional and unconditional probabilities of an event involving only X, conditioned over an event involving only Y. It is evident that $\phi(X|Y)$ measures the degree of interdependence between X and Y. Thus, unlike with mutual information, if $\phi(X|Y) < \phi(W|Z)$, then it can indeed be said that X depends less on Y than W does on Z.

It is easy to show that the ϕ-mixing coefficient has the following properties:

1. $\phi(X|Y) \in [0, 1]$.
2. In general, $\phi(X|Y) \neq \phi(Y|X)$. Thus the ϕ-mixing coefficient gives directional information.
3. X and Y are independent random variables if and only if $\phi(X|Y) = 0$, or equivalently $\phi(Y|X) = 0$.
4. The ϕ-mixing coefficient is invariant under any one-to-one transformation of the data. Thus if $f : \mathbb{A} \to \mathbb{C}$, $g : \mathbb{B} \to \mathbb{D}$ are one-to-one and onto maps, then

$$\phi(X|Y) = \phi(f(X)|g(Y)).$$

In particular, if $\mathbb{A}$, $\mathbb{B}$ are subsets of $\mathbb{R}$, and f, $g : \mathbb{R} \to \mathbb{R}$ are monotonic functions, then once again the above equation holds. This is a very useful when dealing with transformed, as opposed to raw, data.

Next we discuss some methods for either estimating or computing exactly the ϕ-mixing coefficient between two discrete random variables. For each integer n, let $\mathbb{S}_n$ denote the n-dimensional simplex. Thus

$$\mathbb{S}_n := \{\mathbf{v} \in \mathbb{R}^n : v_i \geq 0 \ \forall i, \sum_{i=1}^{n} v_i = 1\}.$$

If $\mathbb{A} = \{a_1, \ldots, a_n\}$ and $\boldsymbol{\mu} \in \mathbb{S}_n$, then $\boldsymbol{\mu}$ defines a measure P_μ on the set $\mathbb{A}$ according to

$$P_\mu(S) = \sum_{i=1}^{n} \mu_i I_S(a_i),$$

where $I_S(\cdot)$ denotes the indicator function of S. To avoid more notation, we will write $\mu(S)$ instead of the more precise $P_\mu(S)$.

Now suppose $\mathbb{A}$, $\mathbb{B}$ denotes sets of cardinality n, m respectively, and that $\boldsymbol{\mu} \in \mathbb{S}_n$, $\boldsymbol{\nu} \in \mathbb{S}_m$ are the marginal distributions of random variables X, Y assuming values in $\mathbb{A}$, $\mathbb{B}$ respectively. Then the distribution $\psi \in \mathbb{S}_{nm}$ defined by $\psi_{ij} = \mu_i \nu_j$ is called the **product distribution** on $\mathbb{A} \times \mathbb{B}$. It is the distribution that the pair (X, Y) would have if X, Y were independent. In the other direction, if $\theta \in \mathbb{S}_{nm}$ is a distribution on $\mathbb{A} \times \mathbb{B}$, representing the joint distribution of (X, Y), then $\theta_\mathbb{A} \in \mathbb{S}_n$, $\theta_\mathbb{B} \in \mathbb{S}_m$ defined respectively by

$$(\theta_\mathbb{A})_i := \sum_{j=1}^{m} \theta_{ij}, \ (\theta_\mathbb{B})_j := \sum_{i=1}^{n} \theta_{ij}$$

are called the **marginal distributions** of θ on $\mathbb{A}$ and $\mathbb{B}$ respectively. These are the distributions of X by itself and of Y by itself, respectively. With this notation, the definition of the ϕ-mixing coefficient can be stated equivalently as follows:

$$\phi(X|Y) := \max_{S \subseteq \mathbb{A}, T \subseteq \mathbb{B}} \left| \frac{\theta(S \times T)}{\nu(T)} - \mu(S) \right|. \tag{3.22}$$

If we use (3.21) directly to compute $\phi(X|Y)$, then it would be necessary to enumerate all possible subsets of $\mathbb{A}, \mathbb{B}$, which would involve 2^{n+m} operations. It is now shown that there exist readily computable upper and lower bounds for $\phi(X|Y)$. Moreover, in the special case where ν is the uniform distribution, it is possible to arrive at an exact value for $\phi(X|Y)$. For this purpose we recall the definition of the matrix induced norm. Let us define $\psi = \mu \times \mu$ to be the product distribution of the two marginals, and define

$$\lambda_{ij} := \theta_{ij} - \psi_{ij}, \Lambda := [\lambda_{ij}] \in [-1, 1]^{n \times m}.$$

For indices i and j, let $\boldsymbol{\lambda}^i, \boldsymbol{\lambda}_j$ denote respectively the i-th row and j-th column of the matrix Λ. The quantity

$$\|\Lambda\|_{i1} := \max_{1 \leq j \leq m} \sum_{i=1}^{n} |\lambda_{ij}| = \max_{1 \leq j \leq m} \|\boldsymbol{\lambda}_j\|_1$$

is called the ℓ_1-**induced matrix norm** of Λ. It is well-known that

$$\|\Lambda\|_{i1} = \max_{\|\mathbf{v}\|_1 \leq 1} \|\Lambda \mathbf{v}\|_1 = \max_{\mathbf{v} \neq \mathbf{0}} \frac{\|\Lambda \mathbf{v}\|_1}{\|\mathbf{v}\|_1}.$$

With this notation we are ready to state the main results of this section.

Theorem 3.4. *We have that*

$$\frac{0.5\|\Lambda\|_{i1}}{\max_j \nu_j} \leq \phi(X|Y) \leq \frac{0.5\|\Lambda\|_{i1}}{\min_j \nu_j}. \tag{3.23}$$

In particular, if ν is the uniform distribution on $\mathbb{B}$, then

$$\phi(X|Y) = 0.5m\|\Lambda\|_{i1}. \tag{3.24}$$

Proof. The first step is to get rid of the absolute value sign in the definition of the ϕ-mixing coefficient, by showing that

$$\phi(X|Y) = \max_{S \subseteq \mathbb{A}, T \subseteq \mathbb{B}} \left[\frac{\theta(S \times T)}{\nu(T)} - \mu(S) \right]. \tag{3.25}$$

For this purpose, define

$$\mathcal{R}_\phi := \left\{ \frac{\theta(S \times T)}{\nu(T)} - \mu(S), S \subseteq \mathbb{A}, T \subseteq \mathbb{B} \right\}.$$

Then $\mathcal{R}_\phi$ is a subset of the real line consisting of at most 2^{n+m} elements. Now it is claimed that the set $\mathcal{R}_\phi$ is symmetric; that is, $x \in \mathcal{R}_\phi$ implies that $-x \in \mathcal{R}_\phi$. If this claim can be established, then (3.25) follows readily. So suppose $x \in \mathcal{R}_\phi$, and choose $S \subseteq \mathbb{A}$, $T \subseteq \mathbb{B}$ such that

$$\frac{\theta(S \times T)}{\mu(T)} - \nu(S) = x, \text{ or } \theta(S \times T) - \mu(S)\nu(T) = x\nu(T).$$

Let S^c denote the complement of S in $\mathbb{A}$. Then, using the facts that

$$\mu(S^c) = 1 - \mu(S),$$

$$\theta(S^c \times T) = \theta(\mathbb{A} \times T) - \theta(S \times T) = \nu(T) - \theta(S \times T),$$

it is easy to verify that

$$\theta(S^c \times T) - \mu(S^c)\nu(T) = -x\nu(T).$$

So $\mathcal{R}_\phi$ is symmetric and (3.25) follows.

To facilitate the proof the theorem, we introduce a map from the power set of $\mathbb{A}$ into $\{0, 1\}^n$. For a subset $S \subseteq \mathbb{A}$, we define $\mathbf{h}(S) \in \{0, 1\}^n$ by

$$h_i(S) = \begin{cases} 1, & \text{if } a_i \in S, \\ 0, & \text{if } a_i \notin S. \end{cases}$$

The map $\mathbf{h} : 2^\mathbb{B} \rightarrow \{0, 1\}^m$ is defined analogously. With these definitions, it is obvious that, for $S \subseteq \mathbb{A}$, $T \subseteq \mathbb{B}$, we have

$$\mu(S) = [\mathbf{h}(S)]^t \mu = \mu^t \mathbf{h}(S), \nu(T) = [\mathbf{h}(T)]^t \nu = \nu^t \mathbf{h}(T),$$

$$\theta(S \times T) = [\mathbf{h}(S)]^t \Theta \mathbf{h}(T),$$

where $\Theta = [\theta_{ij}]$. By replacing $\mathbf{h}(S)$ and $\mathbf{h}(T)$ by arbitrary binary vectors $\mathbf{a} \in \{0, 1\}^n$, $\mathbf{b} \in \{0, 1\}^m$, it readily follows from (3.25) that

$$\phi(X|Y) = \max_{\mathbf{a} \in \{0,1\}^n, \mathbf{b} \in \{0,1\}^m} \frac{\mathbf{a}^t \Lambda \mathbf{b}}{\nu^t \mathbf{b}}. \tag{3.26}$$

Next, it is clear that

$$\max_{\mathbf{a} \in \{0,1\}^n, \mathbf{b} \in \{0,1\}^m} \frac{\mathbf{a}^t \Lambda \mathbf{b}}{\nu^t \mathbf{b}} = \max_{\mathbf{b} \in \{0,1\}^m} \max_{\mathbf{a} \in \{0,1\}^n} \frac{\mathbf{a}^t \Lambda \mathbf{b}}{\nu^t \mathbf{b}}.$$

Now rewrite

$$\frac{\mathbf{a}^t \Lambda \mathbf{b}}{\boldsymbol{\nu}^t \mathbf{b}} = \sum_{i=1}^{n} a_i \frac{\boldsymbol{\lambda}^i \mathbf{b}}{\boldsymbol{\nu}^t \mathbf{b}}.$$

Therefore, for a fixed $\mathbf{b} \in \{0, 1\}^m$, the inner maximum is achieved by the choice

$$a_i = \begin{cases} 1, & \text{if } (\boldsymbol{\lambda}^i \mathbf{b})/(\boldsymbol{\nu}^t \mathbf{b}) \geq 0, \\ 0, & \text{if } (\boldsymbol{\lambda}^i \mathbf{b})/(\boldsymbol{\nu}^t \mathbf{b}) < 0. \end{cases},$$

and

$$\max_{\mathbf{a} \in \{0,1\}^n} \frac{\mathbf{a}^t \Lambda \mathbf{b}}{\boldsymbol{\nu}^t \mathbf{b}} = \sum_{i=1}^{n} \left(\frac{\boldsymbol{\lambda}^i \mathbf{b}}{\boldsymbol{\nu}^t \mathbf{b}} \right)_+.$$

As a result, we have now an alternate formula for $\phi(X|Y)$, namely

$$\phi(X|Y) = \max_{\mathbf{b} \in \{0,1\}^m} \sum_{i=1}^{n} \left(\frac{\boldsymbol{\lambda}^i \mathbf{b}}{\boldsymbol{\nu}^t \mathbf{b}} \right)_+. \qquad (3.27)$$

Next, let $\mathbf{e}$ denote a column vector consisting of all ones, with the subscript denoting its dimension, and observe that

$$\boldsymbol{\mu}^t = \mathbf{e}_n^t \Theta = \mathbf{e}_n^t \Psi \implies \mathbf{e}_n^t \Lambda = \mathbf{0}_n, \text{ similarly } \Lambda \mathbf{e}_m = \mathbf{0}_m.$$

Therefore, for any vector $\mathbf{v} \in \mathbb{R}^m$, it follows that

$$\mathbf{e}_n^t \Lambda \mathbf{v} = 0 \implies \sum_{i=1}^{n} \boldsymbol{\lambda}^i \mathbf{v} = 0$$

$$\implies \sum_{i=1}^{n} (\boldsymbol{\lambda}^i \mathbf{v})_+ + \sum_{i=1}^{n} (\boldsymbol{\lambda}^i \mathbf{v})_- = 0$$

$$\implies \sum_{i=1}^{n} (\boldsymbol{\lambda}^i \mathbf{v})_+ = - \sum_{i=1}^{n} (\boldsymbol{\lambda}^i \mathbf{v})_-$$

$$\implies \sum_{i=1}^{n} (\boldsymbol{\lambda}^i \mathbf{v})_+ = 0.5 \sum_{i=1}^{n} |\boldsymbol{\lambda}^i \mathbf{v}| = 0.5 \|\Lambda \mathbf{v}\|_1. \qquad (3.28)$$

So in particular it follows that

$$\sum_{i=1}^{n} \left(\frac{\boldsymbol{\lambda}^i \mathbf{b}}{\boldsymbol{\nu}^t \mathbf{b}} \right)_+ = 0.5 \frac{\|\Lambda \mathbf{b}\|_1}{\boldsymbol{\nu}^t \mathbf{b}}.$$

To prove the lower bound in (3.23), choose an index j_0 such that $\|\boldsymbol{\lambda}_{j_0}\|_1 = \|\Lambda\|_{i1}$, and choose $\mathbf{b}_0 \in \{0, 1\}^m$ such that $b_{j_0} = 1$ and $b_j = 0$ for all $j \neq j_0$. Then

$$\sum_{i=1}^{n}\left(\frac{\boldsymbol{\lambda}^i \mathbf{b}_0}{\boldsymbol{\nu}\mathbf{b}_0}\right)_+ = \frac{1}{\nu_{j_0}}\sum_{i=1}^{n}(\lambda_{i,j_0})_+ = \frac{0.5}{\nu_{j_0}}\sum_{i=1}^{n}|\lambda_{i,j_0}| = \frac{0.5\|\Lambda\|_{i1}}{\nu_{j_0}}$$

$$\geq \frac{0.5\|\Lambda\|_{i1}}{\max_j \nu_j}.$$

To prove the upper bound in (3.23), note that for all $\mathbf{b} \in \{0, 1\}^m$, we have

$$\sum_{i=1}^{n}\frac{(\boldsymbol{\lambda}^i\mathbf{b})_+}{\boldsymbol{\nu}^t\mathbf{b}} = 0.5\sum_{i=1}^{n}\frac{|\boldsymbol{\lambda}^i\mathbf{b}|}{\boldsymbol{\nu}^t\mathbf{b}} = 0.5\frac{\|\Lambda\mathbf{b}\|_1}{\boldsymbol{\nu}^t\mathbf{b}}.$$

Now we change the variable of optimization from $\mathbf{b}$ to $\mathbf{v} := \mathrm{Diag}(\boldsymbol{\nu})\mathbf{b}$, and use the fact that induced matrix norm $\|\cdot\|_{i1}$ is submultiplicative. This leads to

$$\phi(X|Y) = 0.5\max_{\mathbf{b}\in\{0,1\}^m}\frac{\|\Lambda\mathbf{b}\|_1}{\boldsymbol{\nu}\mathbf{b}} \leq 0.5\max_{\mathbf{b}\in\mathbb{R}^m}\frac{\|\Lambda\mathbf{b}\|_1}{|\boldsymbol{\nu}\mathbf{b}|}$$

$$= 0.5\max_{\mathbf{v}\in\mathbb{R}^m}\frac{\|\Lambda[\mathrm{Diag}(\boldsymbol{\nu})]^{-1}\mathbf{v}\|_1}{\|\mathbf{v}\|_1} = 0.5\|\Lambda[\mathrm{Diag}(\boldsymbol{\nu})]^{-1}\|_{i1}$$

$$\leq 0.5\|\Lambda\|_{i1}\cdot\|[\mathrm{Diag}(\boldsymbol{\nu})]^{-1}\|_{i1} = \frac{0.5\|\Lambda\|_{i1}}{\min_j \nu_j}.$$

Finally, if $\boldsymbol{\nu}$ is the uniform distribution, then $\min_j \nu_j = \max_j \nu_j = 1/m$. So the two inequalities in (3.23) become equalities. $\qquad\square$

3.6.2 Data Processing Inequality for the Phi-Mixing Coefficient

In this section we state and prove an analog of the data processing inequality (3.11) for the ϕ-mixing coefficient.

Theorem 3.5. *Suppose* $(X \perp Z)|Y$. *Then*

$$\phi(X|Z) \leq \min\{\phi(X|Y), \phi(Y|Z)\}, \tag{3.29}$$

$$\phi(Z|X) \leq \min\{\phi(Z|Y), \phi(Y|X)\}. \tag{3.30}$$

Proof. We begin by restating (3.27) in terms of subsets of $\mathbb{B}$. Recall from (3.27) that

$$\phi(X|Y) = \max_{\mathbf{b}\in\{0,1\}^m}\sum_{i=1}^{n}\left(\frac{\boldsymbol{\lambda}^i\mathbf{b}}{\boldsymbol{\nu}^t\mathbf{b}}\right)_+.$$

Again, recall that $\Lambda = \Theta - \Psi$, so that $\boldsymbol{\lambda}^i = \boldsymbol{\theta}^i - \mu_i\boldsymbol{\nu}^t$. Therefore the above equation becomes

$$\phi(X|Y) = \max_{\mathbf{b} \in \{0,1\}^m} \sum_{i=1}^{n} \left(\frac{\theta^t \mathbf{b}}{\nu^t \mathbf{b}} - \mu_i \right)_+ .$$

This formula can be restated in terms of probabilities, as follows:

$$\phi(X|Y) = \max_{T \subseteq \mathbb{B}} \sum_{i=1}^{n} [\Pr\{X = i | Y \in T\} - \Pr\{X = i\}]_+. \qquad (3.31)$$

Suppose $(X \perp Z)|Y$. Since the ϕ-mixing coefficient is not symmetric, it is necessary to prove two distinct inequalities, namely: (i) $\phi(X|Z) \leq \phi(X|Y)$, and (ii) $\phi(X|Z) \leq \phi(Y|Z)$.

Proof that $\phi(X|Z) \leq \phi(X|Y)$: For $S \subseteq \mathbb{A}$, define

$$r_\phi(S) := \max_{T \subseteq \mathbb{B}} \Pr\{X \in S | Y \in T\},$$

and observe that

$$\phi(X|Y) = \max_{S \subseteq \mathbb{A}} [r_\phi(S) - \boldsymbol{\mu}(S)].$$

For a given $S \subseteq \mathbb{A}$, choose $T^* = T^*(S) \subseteq \mathbb{B}$ such that

$$\Pr\{X \in S | Y \in T^*\} = r_\phi(S).$$

Suppose $U \subseteq \mathbb{C}$ is arbitrary. Then

$$\Pr\{X \in S \& Z \in U\} = \sum_{j=1}^{m} \Pr\{X \in S \& Y = j \& Z \in U\}$$

$$= \sum_{j=1}^{m} \Pr\{X \in S | Y = j\} \Pr\{Z \in U | Y = j\} \Pr\{Y = j\}$$

$$= \sum_{j=1}^{m} \Pr\{X \in S | Y = j\} \Pr\{Z \in U \& Y = j\}$$

$$\leq r_\phi(S) \sum_{j=1}^{m} \Pr\{Z \in U \& Y = j\}$$

$$= r_\phi(S) \Pr\{Z \in U\}.$$

Dividing both sides by $\Pr\{Z \in U\}$ leads to

$$\Pr\{X \in S | Z \in U\} \leq r_\phi(S),$$

$$\Pr\{X \in S | Z \in U\} - \mu(S) \leq r_\phi(S) - \mu(S) \leq \phi(X|Y).$$

Proof that $\phi(X|Z) \leq \phi(Y|Z)$: Let us define

$$c(S, U) := \Pr\{X \in S | Z \in U\} - \mu(S),$$

and reason as follows:

$$
\begin{aligned}
c(S, U) &= \Pr\{X \in S | Z \in U\} - \Pr\{X \in S\} \\
&= \sum_{j=1}^{m} [\Pr\{X \in S \& Y = j | Z \in U\} - \Pr\{X \in S \& Y = j\}] \\
&= \sum_{j=1}^{m} [\Pr\{X \in S | Y = j \& Z \in U\} \Pr\{Y = j | Z \in U\} - \Pr\{X \in S | Y = j\} \Pr\{Y = j\}] \\
&= \sum_{j=1}^{m} \Pr\{X \in S | Y = j\} [\Pr\{Y = j | Z \in U\} - \Pr\{Y = j\}] \\
&\leq \sum_{j=1}^{m} \Pr\{X \in S | Y = j\} [\Pr\{Y = j | Z \in U\} - \Pr\{Y = j\}]_{+} \\
&\leq \sum_{j=1}^{m} [\Pr\{Y = j | Z \in U\} - \Pr\{Y = j\}]_{+} \\
&\leq \max_{U \subseteq \mathcal{C}} \sum_{j=1}^{m} [\Pr\{Y = j \& Z \in U\} - \Pr\{Y = j\}]_{+} \\
&= \phi(Y|Z).
\end{aligned}
$$

Since the right side is independent of both S and U, the desired conclusion follows. $\qquad\square$

3.6.3 A New Algorithm for Inferring GINs

In this section we present a detailed description of our new algorithm for inferring gene interaction networks from expression data. Since it is based on computing the ϕ-mixing coefficient between each pair of genes, the algorithm is referred to as the ϕ-xer algorithm. Recall that there are n genes and m samples of each gene, and the objective of the exercise is to infer whether or not $(X_i \perp X_k)|X_j$ for each pairwise distinct triplet (i, j, k). We begin with the theory behind the algorithm in the case where we know the exact value of the coefficient $\phi(X_i|X_j)$ for each pair of indices i, j. Then we discuss how the algorithm can be implemented in practice, taking account the fact that we can only estimate the coefficient based on a finite number of samples.

So let us begin by assuming (somewhat unrealistically) that exact values are available for all $n(n-1)$ coefficients $\phi(X_i|X_j)$ for each pair of indices $i, j, i \neq j$. Then we proceed as follows: Start with a complete graph of n nodes, where there is a directed edge between every pair of distinct nodes ($n(n-1)$ edges). For each triplet i, j, k of pairwise distinct indices, check whether the DPI-like inequality

$$\phi(X_i|X_k) \leq \min\{\phi(X_i|X_j), \phi(X_j|X_k)\} \tag{3.32}$$

holds. If so, discard the edge from node k to node i, but retain a 'phantom' edge for future comparison purposes.

This step is referred to as 'pruning'. Note that the pruning operation can at best replace a direct path of length one (i.e. an edge) by an indirect path of length two. However, one or both of those edges could be 'phantom' edges that have been pruned in an earlier step. In such a case, there would still exist another path between the two nodes of the phantom edge, though possibly consisting of phantom edges. The argument can be repeated until all phantom edges if any are replaced by real edges (i.e., those that have survived the pruning). Hence the graph that results from the pruning operation is still strongly connected. Also, since any discarded edges are still retained for the purposes of future comparisons, it is clear that the order in which the triplets are processed does not affect the final answer. Note that the complexity of this operation is cubic in n.

At this stage, one can ask whether the graph resulting from the pruning operation has any significance. It is now shown, by invoking the Occam's razor principle (giving the simplest possible explanation), that the graph resulting from pruning is *a minimal graph* consistent with the data set. For this purpose, we define a partial ordering on the set of directed graphs with n nodes whereby $\mathcal{G}_1 \leq \mathcal{G}_2$ if $\mathcal{G}_1$ is a subgraph of $\mathcal{G}_2$, ignoring weights of the edges. For a given triplet i, j, k, it is obvious that $(X_i \perp X_k)|X_j$ if and only if every directed path from node i to node k passes through node j, and also every directed path from node k to node i passes through node j. Now, it follows from the DPI that if $(X_i \perp X_k)|X_j$, then (3.32) holds. Taking the contrapositive shows that if (3.32) is false, then $(X_i \not\perp X_k)|X_j$. Consequently, if (3.32) is false, then there must exist a path from node i to node k that does not pass through node j. Given the sequential nature of the pruning algorithm, when (3.32) is checked for a specific triplet (i, j, k), there already exist edges from node i to node j and from node j to node k; that is, there exists a path of length two from node i to node k. Now, if (3.32) is false, then there must exist another path from node i to node k that does not pass through node j. It is of course possible that this path consists of many edges. However, by the Occam's razor principle, the simplest explanation would be that there is a shorter path of length one, i.e. a directed edge from node k to node i.

What has been shown is that, under the Occam's razor principle, the graph that results from pruning is *minimal* in the following sense. First, it is consistent with the ϕ-mixing coefficients, and second, any other graph that is 'less than' this graph in the partial ordering defined above would not be consistent with the ϕ-mixing coefficients. Thus, if any edges are removed from the graph that results from applying the pruning step, then some other edges would have to be added in order for the graph to be consistent with the ϕ-mixing coefficients. Note that we are obliged to say *a* and not *the* minimal graph, because there might not be a unique minimal graph.

Nevertheless, it is obvious that the application of the algorithm will result in a unique graph, irrespective of the order in which all the triplets (i, j, k) are examined.

Now that the basic theory is in place, it is possible to present an algorithm for reverse-engineering a GIN from experimental expression data. The main issue here is that we do not know the 'true' coefficient $\phi(X_i|X_j)$ exactly. Even if we were to discretize the random variables by binning and then use (3.23), the resulting quantity would still only be an approximation of $\phi(X_i|X_j)$ and not the exact value. Moreover, a direct application of (3.21) would be too expensive computationally. These are some of the considerations that enter into the implementation described below. Recall that there are n genes and m samples of each.

Binning the expression values of each gene: Choose an integer k such that $k \leq \lfloor (m/3)^{1/2} \rfloor$. For each index i, divide the total range of X_i into k bins that correspond to 'percentiles'. Note that percentile binning is also referred to as 'data-dependent partitioning' in [33]. For each pair of indices i, j, and each sample label l, assign the sample pair (x_{il}, x_{jl}) to its associated bin. The discretization ensures that each random variable X_i assumes one of just k values (corresponding to the bins). The choice of k ensures that on average there will be at least three entries in each of the k^2 bins of the joint random variable (X_i, X_j) for each pair (i, j). The choice of percentile discretization (as opposed to, for example, uniformly gridding the range), ensures that the marginal distribution of each X_i is nearly equal to the uniform distribution on k labels, and allows us to use the estimates (3.23). If m is an exact multiple of k then each marginal distribution would indeed be the uniform distribution, and we would have an exact value of the ϕ-mixing coefficient between the discretized random variables. But in general m might not be an exact multiple of k. Percentile binning also ensures that the joint distribution of the discretized pairs (X_i, X_j) remains invariant under any monotonic transformation of the data. It is important to note here that the invariance property holds even if different monotone transformations are applied to different expression variables.

Estimating the ϕ-mixing coefficient: After binning, for each pair of indices i, j, we determine the associated joint distribution of the discretized random variables, which will be a $k \times k$ matrix. For each pair of indices i, j, we use (3.23) to compute an interval $[\phi_l(X_i|X_j), \phi_u(X_i|X_j)]$ that contains the true value $\phi(X_i|X_j)$. Define $\phi_a(X_i|X_j) = [\phi_l(X_i|X_j) + \phi_u(X_i|X_j)]/2$ to be the midpoint of this bounding interval. Note that we are being a bit imprecise since X_i now represents the discretized and not the original (continuous) expression value. However, in the interests of notational simplicity, we ignore this distinction. The complexity of this operation is quadratic in n, the total number of genes.

Pruning: As before, start with a complete graph on n nodes, and then apply the data processing inequality to do the pruning, for each triplet (i, j, k). Since we have only empirically determined values of the mixing coefficient, there are several possible ways of interpreting the data processing inequality. At this stage three different ways of implementing the pruning operation were examined.

1. Eliminate the edge from node j to node i if

$$\phi_u(X_i|X_k) \leq \min\{\phi_l(X_i|X_j), \phi_l(X_j|X_k)\}. \tag{3.33}$$

2. Eliminate the edge from node j to node i if

$$\phi_u(X_i|X_k) \leq \min\{\phi_a(X_i|X_j), \phi_a(X_j|X_k)\}. \tag{3.34}$$

3. Eliminate the edge from node j to node i if

$$\phi_a(X_i|X_k) \leq \min\{\phi_a(X_i|X_j), \phi_a(X_j|X_k)\}. \tag{3.35}$$

Since it is always the case that

$$\phi_l(X_i|X_j) \leq \phi_a(X_i|X_j) \leq \phi_u(X_i|X_j),$$

it is easy to see that any edge that gets pruned out under Rule 1 also gets pruned out under Rule 2, but Rule 2 could also prune out other edges that survive Rule 1. Similar remarks apply to Rule 2 vis-à-vis Rule 3. Thus if let $\mathcal{G}_1, \mathcal{G}_2, \mathcal{G}_3$ denote the graphs produced by applying Rule 1, Rule 2, and Rule 3 to the same data set, then it is easy to see that $\mathcal{G}_3$ is a subgraph of $\mathcal{G}_2$, which is in turn a subgraph of $\mathcal{G}_1$. Based on several numerical experiments, we finally opted to use Rule 2, but with 'thresholding' as explained next.

Thresholding: We have constructed several genome-wide networks from expression data, as detailed in the next section. Our numerical experiments have shown that after the pruning step described above, the GINs that result are characterized by the property that the mean value of all the edge weights is noticeably higher than the median. This means that there are relatively far more edges with low weights than edges with high weights, but the high-weight edges have significantly higher weights. One of the main reasons for reverse-engineering GINs is to identify 'hubs', that is, genes that are connected to many other genes. Again, the GINs that result from the pruning step do not show sufficient variation between the largest node-degree and the smallest node-degree. In 'validating' the reverse-engineered GIN, it is highly desirable to eliminate all of these low-weight edges, while still ensuring that the graph remains strongly connected. Several numerical experiments have suggested the following strategy: After the pruning step, compute the mean μ and standard deviation σ of the weights of all edges. Then eliminate all edges whose weights are below μ. If the thresholded graph is strongly connected, then keep it; otherwise lower the threshold to $\mu - \sigma$, or use no threshold at all. Thus far, in our various experiments, in about half of the cases the threshold of μ has resulted in a strongly connected network, while in one case even the higher threshold of $\mu + \sigma$ has resulted in a strongly connected network. Interestingly, the highest degree nodes in the pruned network still remain the highest degree nodes in the pruned and thresholded network. The numerical examples in later sections make this point clear.

Table 3.1 Lung cancer networks pruned using rule 2

No.	No. of genes (n)	No. of samples (m)	No. of edges and SCCs under various thresholds		
			0	μ	SCCs
1.	19,579	148	3,853,936	1,791,624	1
2.	19,579	108	4,474,834	2,283,114[†]	2
3.	19,579	29	2,548,501	940,785[†]	6

3.6.4 A Case Study: A Lung Cancer GIN

Using the algorithm presented here, we have reverse-engineered several GINs. Due to space and time limitations, we are reporting here only the three GINs obtained from lung cancer cell line data. We have also constructed three different GINs from ovarian cancer tumor tissue data, and three different GINs from melanoma, of which two are based on data from tumor tissues and one is based on data from cell lines. The broad conclusions given below apply to those GINs as well.

The three lung cancer GINs are based on gene expression data from lung cancer cell lines from the laboratory of Prof. John Minna and were provided to us by his student, Alex Augustyn. There were 148 cell lines in all, consisting of 108 non-small cell lung cancer (NSCLC), 11 neuro-endocrine non-small cell lung cancer (NE-NSCLC), and 29 small-cell lung cancer (SCLC). Network 1 is based on combining the data from all these lines, whence the number of samples is 148. Network 2 is based on the NSCLC samples alone, while Network 3 is based on the SCLC samples alone. The objective of this exercise, in the long run, can be explained briefly as follows: SCLC is often associated with smoking, and patients with SCLC have very poor prognosis. In contrast, NSCLC is found in persons who have never smoked nor been exposed to smoking, but the prognosis of patients with NSCLC is significantly better than for those with SCLC. On the other hand, there is a variety of NSCLC, namely NE-NSCLC, for which the prognosis is as poor as for SCLC patients. Thus eventually we would like to be able to understand *why* this is so. This issue is revisited again in Chap. 4.

The raw statistics of the various networks obtained using Rule 2 for pruning are given in Table 3.1. Statistics for the number of edges that remain after thresholding the resulting network with the threshold μ, as well as the number of strongly connected components (SCCs) are also given. Note that the network with no threshold is always strongly connected so the number of SCCs is always one in that case, and is therefore not displayed. The superscript [†] indicates that the resulting thresholded network is *not* strongly connected, i.e. that the threshold is too high.

It can be seen that the networks resulting from pruning using Rule 2 and then thresholding at the level of μ are either strongly connected, or else have a very small number of SCCs, meaning that 'for all practical purposes' they are strongly connected. If we use Rule 1 for pruning, the resulting networks contain far too many edges to be of any use, whereas if we use Rule 3 for pruning, the networks tend to become disconnected into a large number of SCCs at a threshold of μ. For this reason, we have chosen to use Rule 2 for pruning and a threshold of μ for all future analyses.

From these networks, a few general features of our algorithm emerge.

- Recall that the algorithm begins with a complete directed graph on n nodes; therefore the initial number of edges is $n(n-1) \approx n^2$. Applying the pruning step retains the strong connectivity property, as pointed out earlier. In all of the cases studied here, the pruning step eliminates 99 % or more of these edges. Therefore the algorithm is quite efficient in terms of finding a very small GIN consistent with the data.

- Though the network contains a very small number of edges in comparison to the theoretical maximum of $n(n-1)/2$, the networks are nevertheless very shallow, in the sense that from any one gene, it is possible to reach every other gene in three or fewer hops. This is a limitation of the modeling methodology, because it strives for a *logically minimal* representation of the data, whereas there is no reason to suppose that 'real' biological networks are in fact so economical. In other words, it appears that metabolic pathways that consist of several steps in biology are captured by just one edge in the GIN representation. This raises the question of 'harmonizing' the expression data (which is context-specific) with prior knowledge in the form of known pathways. This problem is discussed in Sect. 4.2.

- It can be seen that in two out of three cases, the network thresholded with the mean edge weight μ has fewer than half of the edges in the pruned network, while in the case of the third, this is nearly so. This implies that the mean of the edge weights of the post-pruning network is higher than the median, or equivalently, the pruned network (before thresholding) has a large number of low-weight edges and relatively fewer high-weight edges.

Now we study the 'power law' nature of the degree distributions GIN No. 1, for lung cancer, with the threshold set equal to μ. Note that the validation step discussed subsequently is based on this network.

It is widely believed by biologists that real GINs consist of a few master regulators that control many hundreds of other genes. It is also proposed by some authors that biological GINs show a power law behavior. Specifically, let d denote an integer corresponding to the degree of a node. (For each node, the phrases in-degree, out-degree and total degree are self-explanatory.) Let $n(d)$ denote the number of nodes in the GIN that have degree d. Then the belief is that $n(d)$ asymptotically looks like $d^{-\alpha}$ for some index α. We wished to examine whether this is indeed true for Network No. 1.

It turns out that the total degrees of various nodes does not show much variation; the maximum is 1,735 for ATCAY and the minimum degree is 64. Similarly, the in-degree also does not show much variation, ranging from 411 to 34. In contrast, the out-degree, that is to say, the number of downstream neighbors of a gene, varies from a high of 1,626 to a low of 1, which is the theoretical minimum.

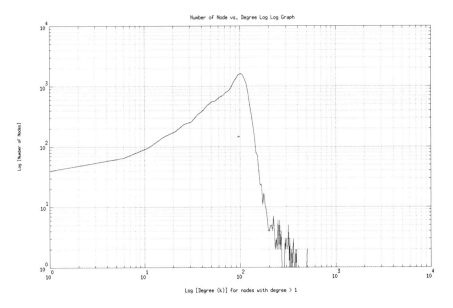

Fig. 3.4 Plot of degree versus number of nodes all degrees

(Note that if any node had in- or out-degree of zero, then the network would not be strongly connected.) For this reason, we studied the distribution of $n(d)$ as a function of the out-degree d, and plotted this on a log-log scale. To avoid the graph becoming too jerky, we 'binned' the degree d into bins of width five. That is to say, we computed the number of nodes with degrees between 1 and 5, and then between 6 and 10, and so on. The graph for the entire range of degrees is not particularly informative and does not show any power law behavior. However, we observed that when the node degree is between 100 and 300, the graph did seem to show a linear relationship (between $\log d$ and $\log n(d)$). Therefore we zoomed in on the plot for degrees lying between 100 and 300, and this time we could observe a fairly clear power law behavior with exponent of -6.88 (Fig. 3.4).

It is virtually impossible to validate an entire GIN since there is no known and confirmed 'absolute truth' against which predictions can be confirmed. Existing examples of GINs such as [2–5] are often obtained by combining small interaction networks from a variety of sources. The difficulty with this approach is that *the context* in which these small interaction networks are determined need not be the same across all of them. Hence combining these small individual networks into one large patchwork network cannot be justified biologically, and the resulting network cannot be accepted as reflecting 'reality'. It must be emphasized that the entire *raison d'être* of our algorithm is to embrace the entire genome (or as much of it as is covered by the data) to produce a *context-specific, genome-wide network*. Hence it would be inappropriate to compare the reverse-engineered networks with the networks available in public domain databases. In any case, even the largest networks

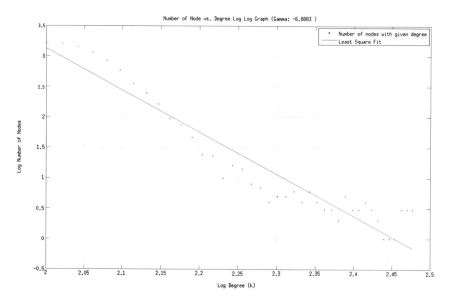

Fig. 3.5 Plot of degree versus number of nodes degrees between 100 and 300

obtained in this (somewhat dubious) manner still cover only about half of the 22,000 or so human genes; see [5]. In contrast, the networks that we have reverse-engineered routinely combine *all* genes studied in the experiment, of the order of 20,000, as seen from the previous section (Fig. 3.5).

This therefore raises the question of just how such reverse-engineered networks are to be validated. After considerable thought, we chose to use evidence from so-called ChIP-seq tests for a few transcription factors. A transcription factor is a special kind of gene that is involved in regulating the transcription of other genes.[8] ChIP-seq stands for 'chromatin immuno-precipitation sequencing'. A good introduction to ChIP-seq for non-biologists can be found at [35]. The basic idea is that a transcription factor is immuno-precipitated with the entire genome. As a result, several DNA fragments bind to the transcription factor. After these fragments are isolated through other experimental techniques, the fragments are then sequenced. In principle each DNA fragment represents one or more genes that are regulated by the transcription factor. However, the raw experimental technique is rife with false positives and produces literally thousands of genes as being potentially regulated by the transcription factor under study. Further analysis is needed to weed out these false positives, and thus produce a realistic set of potential downstream target genes. For one of the transcription factors studied here, namely ASCL1, our collaborators (Prof. Jane Johnson and Mr. Mark Borromeo) used an algorithm called GREAT (Genomic Regions Enhancement of Annotations Tool) [36] to eliminate most of these false

---

[8] Recall that the 'central dogma' of biology, as enunciated by Francis Crick [34] states that DNA is converted to RNA (transcription) which is then converted to protein(s) (translation).

positives, and produce a set of potential target genes. For the other two transcription factors studied here, namely PPARG and NKX2-1, the laboratory of Prof. Ralf Kittler, specifically Dr. Rahul Kollipara, produced a set of potential target genes for each transcription factor using a different peak-calling routine.

As mentioned above, applying GREAT or another similar algorithm to the raw ChIP-seq data results in a set of potential downstream target genes of the transcription factor under study. This list of genes can then be compared with the downstream neighbors of the same transcription factor in the reverse-engineered GIN. It is also reasonable to consider *all* first-order neighbors of the transcription factor, both up-stream as well as down-stream, and to compare this list against the set of predicted target genes produced by ChIP-seq analysis.[9] The inclusion of both up-stream as well as down-stream neighbors can be justified on the basis that biological networks are full of local feedback loops, whence even up-stream neighbors of a transcription factor can be viewed as being potentially 'regulated' by that transcription factor. In any case, as can be seen below, the results of the validation are not affected very much if we were to consider only first-order downstream neighbors of each transcription factor.

The validation now consists of seeing whether the set of first-order neighbors is 'enriched' for ChIP-seq genes. Let k denote the number of ChIP-seq genes of the transcription factor under study, and as before let n denote the total number of genes in the study. Then the probability that a randomly selected gene is a ChIP-seq gene is $p = k/(n-1)$. If the transcription factor has s neighbors out of which l are ChIP-seq genes, then there is enrichment only if $l/s > p$. To compute the likelihood of this observation occurring purely due to chance, we use a simple binomial model. Moreover, since all l neighbors are distinct, the hypergeometric distribution is the correct one to use. Therefore the likelihood that the enrichment is purely due to chance is given by the `Matlab` command

$$L = 1 - \text{hygecdf}(l-1, n-1, k, s).$$

This number L is the so-called P-value that biologists use to determine whether the observed outcome is due purely to chance.

As mentioned above, we obtained ChIP-seq data for three transcription factors in lung cancer tissues, namely: ASCL1, PPARG, and NKX2-1. In biology, a likelihood (P-value) less than 0.05 is considered significant, and in all cases the computed likelihood is comparable to this threshold. Table 3.2 shows the relative likelihood that a randomly chosen gene is a ChIP gene for each of these transcription factors.

Table 3.3 shows the enrichment of ChIP genes amongst the first-order downstream neighbors, and amongst all neighbors, for these three transcription factors. In this table, an entry of zero for the likelihood means that the number is smaller than machine zero.

---

[9] In the interests of brevity, these are referred to, somewhat inaccurately, as 'ChIP-seq genes'.

Table 3.2 Number of potential target genes for various transcription factors

Item	ASCL1	PPARG	NKX2-1
Total no. of genes	19,579	19,579	19,579
Total no. of ChIP genes	226	221	684
Prob. of being a ChIP gene	0.0115	0.0113	0.0349

Table 3.3 Enrichment of neighbors for ChIP genes

Gene name	Up-neigh.	ChIP genes	L_d	Tot. neigh.	ChIP genes	L_t
ASCL1	690	84	0	766	87	0
PPARG	84	3	0.0696	208	5	0.0872
NKX2-1	114	6	0.2089	244	14	0.0481

3.7 Evaluation and Validation of Competing Approaches

Given that the computational biology literature is full of various approaches for reverse-engineering GINs, there is a lot of interest in assessing the relative performance of all the competing approaches. In the area of protein structure prediction based on the primary structure (i.e., the sequence of amino acids that constitute the protein), there is a well-established biennial competition known as CASP (Critical Assessment of Structure Prediction). In this competition, the organizers first determine the 3-D structure of a protein using x-ray crystallography or some other method, but do not share it with the community at large. Instead the community is challenged to 'predict' the structure, and the ones who come closest to the true structure are recognized as such. Perhaps drawing inspiration from this, the research community working in the area of inferring GINs has a competition called DREAM (Dialog for Reverse Engineering Assessment and Methods). In the personal opinion of the author, the DREAM competition lacks the authenticity of the CASP competition, simply because in CASP there is an unambiguous, objective truth that everyone is striving to find, and against which any and all predictions can be compared. This is definitely not the case in DREAM. Rather, in the case of DREAM, synthetic data is generated using some model or combination of models. It should be clear that, given two algorithms, one can always generate data sets on which one algorithm outperforms the other, and other data sets on which the performance is reversed. Until and unless our knowledge of GINs proceeds to a stage where at least a few GINs are completely identified to constitute 'the truth' (as in CASP and protein structures), there is a danger that such competitions actually serve to confuse rather than to clarify. Again, this is the author's personal opinion.

References

1. Zhou, Q., et al.: A gene regulatory network in mouse embryonic stem cells. Proc. Natl. Acad. Sci. **104**(42), 16,438–16,443 (2007)
2. Intact: http://www.ebi.ac.uk/intact/
3. Mint: http://mint.bio.uniroma2.it/mint/welcome.do
4. Biogrid: http://thebiogrid.org/
5. String: http://thebiogrid.org/
6. Komurov, K., White, M.A., Ram, P.T.: Use of data-biased random walks on graphs for the retrieval of context-specific networks from genomic data. PLoS Comput. Biol. **6**(8), (2010)
7. Szklarczyk, D., et al.: The string database in 2011: functional interaction networks of proteins, globally integrated and scored. Nucleic Acids Res. **39**, D561–D568 (2011)
8. Kim, Y., et al.: Principal network analysis: identification of subnetworks representing major dynamics using gene expression data. Bioinformatics **27**(3), 391–398 (2011)
9. Pe'er, D., Hacohen, M.: Principles and strategies for developing network models in cancer. Cell **144**, 864–873 (2011)
10. Basso, K., et al.: Reverse engineering of regulatory networks in human b cells. Nat. Genet. **37**(4), 382–390 (2005)
11. GEO: http://www.ncbi.nlm.nih.gov/geo/
12. TCGA: http://cancergenome.nih.gov
13. Butte, A.J., Kohane, I.S.: Mutual information relevance networks: functional genomic clustering using pairwise entropy measures. Pac. Symp. Biocomput. **5**, 418–429 (2000)
14. Margolin, A.A., et al.: Aracne: an algorithm for the reconstruction of gene regulatory networks in a cellular context. BMC Bioinform. **7(Supplement 1)**, S7 (2008)
15. Chow, C.K., Liu, C.N.: Approximating discrete probability distributions with dependence trees. EEE Trans. Info. Thy. **14**(3), 462–467 (1968)
16. Cover, T.M., Thomas, J.A.: Elements of Information Theory, 2nd edn. Wiley Interscience, New York (2006)
17. Medterms: http://www.medterms.com
18. Wang, K., et al.: Genome-wide identification of post-translational modulators of transcription factor activity in human b cells. Nat. Biotechnol. **27**(9), 829–839 (2009)
19. Zhao, W., Serpedin, E., Dougherty, E.R.: Inferring connectivity of genetic regulatory networks using information theoretic criteria. IEEE/ACM Trans. Comput. Biol. Bioinf. **5**(2), 262–274 (2008)
20. Sklar, M.: Fonctions de répartition à n dimension et leurs marges. Publications de l'Institut Statistiques, Université de Paris **8**, 229–231 (1959)
21. Durante, F., Sempi, C.: Copula theory: An introduction. In: Jaworski, P., Durante, F., Härdle, W., Rychlik, T. (eds.) Copula Theory and Its Applications. Lecture Notes in Statistics. Springer, Berlin (2010)
22. Qiu, P., Gentles, A.J., Plevritis, S.K.: Reducing the computational complexity of information theoretic approaches for reconstructing gene regulatory networks. J. Comput. Biol. **17**(2), 169–176 (2010)
23. Belcastro, V., et al.: Reverse engineering and analysis of genome-wide gene regulatory networks from gene expression profiles using high-performance computing. IEEE/ACM Trans. Comput. Biol. Bioinform. **9**(3), 668–674 (2012)
24. Pearl, J.: Probabilistic Reasoning in Intelligent Systems. Morgan Kaufmann, San Francisco (1988)
25. Friedman, N., et al.: Using bayesian networks to analyze expression data. J. Comput. Biol. **7**(3–4), 601–620 (2000)
26. Barash, Y., Friedman, N.: Context-specific bayesian clustering for gene expression data. J. Comput. Biol. **9**(2), 169–191 (2002)
27. Friedman, N.: Inferring cellular networks using probabilistic graphical models. Science **303**, 799–805 (2004)

28. Heckerman, D.: A tutorial on learning with bayesian networks. In: Jordan, M.I. (ed.) Learning in Graphical Models. MIT Press, Cambridge (1998)
29. Spitzer, F.: Markov random fields and gibbs ensembles. Am. Math. Monthly **78**(2), 142–154 (1971)
30. Vidyasagar, M.: Learning and Generalization: With Applications to Neural Networks and Control Systems. Springer, London (2003)
31. Doukhan, P.: Mixing: Properties and Examples. Springer, Heidelberg (1994)
32. Ibragimov, I.A.: Some lilmit theorems for stationary processes. Theor. Probab. Appl. **7**, 349–382 (1962)
33. Wang, Q., Kulkarni, S.R., Verdú, S.: Divergence estimation of continuous distributions based on data-dependent partitions. IEEE Trans. Informa. Theor. **51**(9), 3064–3074 (2005)
34. Crick, F.H.C.: Central dogma of molecular biology. Nature **227**, 561–563 (1970)
35. Liu, E.T., Pott, S., Huss, M.: Q&a: Chip-seq technologies and the study of gene regulation. BMC Biol. **8**, 56 (2010). http://www.biomedcentral.com/1741-7007/8/56
36. McLean, C.Y., et al.: Great improves functional interpretation of cis-regulatory regions. Nat. Biotechnol. **28**(5), 495–501 (2010)

Chapter 4
Some Research Directions

Abstract In this final chapter, three different directions for future research are sketched. The first problem is that of harmonizing prior knowledge about gene interaction networks that is scattered throughout the literature with the output of the phixer algorithm. This is formulated as a problem in graph theory, and possible approaches are indicated. The second problem is to identify 'genomic machines', that is, sets of genes that are connected by edges that are all over-expressed, or all under-expressed, in a common context. This problem is formulated as one of computing (or at least approximating) the stationary distribution of a large Markov chain, where the states correspond to individual genes. The last problem is to separate causal mutations (drivers of cancer) from coincidental mutations (passengers in cancer). It is surmised that a seven-dimensional vector known as the developmental gene expression profile plays a role in discriminating between drivers and passengers. Preliminary evidence from colorectal cancer is examined, and it is suggested that further studies should be carried out using recently published comprehensive analysis of colorectal cancer.

Keywords Personal medicine · Markov chains · Stationary distributions · Causal mutations · Coincidental mutations · CAN genes

The manner of working of the biology community is to generate and then validate very specific and focused hypotheses, and use the validated hypotheses to build up 'the big picture'. For instance, methods such as the ϕ-xer algorithm presented in the previous chapter will be appreciated by biologists only if they lead to some explicit hypotheses that can then be tested in a laboratory. This inductive bottom-up approach to learning the truth contrasts sharply with the deductive top-down approach of mathematicians and mathematically minded engineers. With this point in mind, in the present chapter we discuss how the context-specific and genome-wide GINs generated by the ϕ-xer algorithm can be 'queried' so as to generate hypotheses. In particular, we discuss how such networks can potentially be used to carry out a 'process-level' annotation of a network, to complement the 'gene-level' annotation that arises from gene-centric experiments such as ChIP-seq.

M. Vidyasagar, *Computational Cancer Biology*, SpringerBriefs in Control,
Automation and Robotics, DOI: 10.1007/978-1-4471-4751-0_4, © The Author(s) 2012

4.1 Harmonizing Prior Knowledge with Phixer Output GIN

As was pointed out in Sect. 3.6, the networks produced by the ϕ-xer algorithm are very efficient in terms of using fewer than 1% of all the possible edges in order to produce a logically minimal GIN that is consistent with the computed ϕ-mixing coefficients. That is the good part. The bad part is that the networks are invariably 'shallow', in the sense that from key transcription factors it is possible to reach all genes in three or fewer hops. It is widely believed that actual biological pathways are far longer than this. The disparity can be explained by the observation that the ϕ-xer algorithm compresses several biological pathways into a single logical pathway (i.e., an edge). On the other hand, without some sort of prior constraints, the algorithm cannot by itself produce longer paths.

At present there are several examples of GINs both in the public domain and in the realm of commercial products; see for example [1–4]. These networks of often obtained by combining small interaction networks from a variety of sources. As a result, any network that is a compendium of this kind *lacks context*. On the other hand, one could adopt the viewpoint that every edge in such a network represents a 'real' interaction in some context of the other, though perhaps not in the context of interest. The question therefore arises: How is it possible to 'harmonize' the prior information contained in such GINs with the output ϕ-xer?

One possible approach is the following: Suppose we have available a network, representing 'the universe of all known interactions, whatever be the context'. Let us denote this graph as $\mathcal{G}_p$ where the subscript indicates 'prior'. Existing networks in the literature are definitely unweighted, and some of the edges can be undirected as well. In such a case, we would replace an undirected edge by two directed edges. Next we construct a network using the ϕ-xer algorithm, call if $\mathcal{G}_\phi$. In such a situation, every edge in $\mathcal{G}_\phi$ represents a context-specific interaction, though it is possible that the interaction is actually a compression of a longer pathway, whereas every edge in $\mathcal{G}_p$ represents an interaction that is potentially present. By taking the intersection of both graphs, and assigning to each edge in the resulting graph its weight from $\mathcal{G}_\phi$, we arrive at what might be called a 'context-specific scaffold', denoted by $\mathcal{G}_c$. One characteristic of the prior GINs represented by $\mathcal{G}_p$ is that they do not often cover all genes, and even for the genes that are covered, there is not always a path between every pair of genes; in other words, $\mathcal{G}_p$ may not be strongly connected. Since $\mathcal{G}_c$ is a subgraph of $\mathcal{G}_p$, in general $\mathcal{G}_c$ will also fail to be strongly connected. So a 'global' question that one can ask is the following: What is the minimum set of edges that can be added to $\mathcal{G}_c$ so as to make it strongly connected? In answering this question, it appears reasonable to add edges *in order of decreasing weight in the graph* $\mathcal{G}_\phi$, because a higher weight edge connotes a stronger interaction between the genes. So in principle we can sort the edges of $\mathcal{G}_\phi$ by weight, and add them in order of decreasing weight until a strongly connected graph results. In fact one can go further. If the initial graph $\mathcal{G}_c$ is not strongly connected, then it has a unique decomposition into strongly connected components (modulo certain permutation of labels); see [5]. So new edges from $\mathcal{G}_\phi$ should be added in order of decreasing weight but only if adding each edge results in reducing the number of connected components. In this way one will arrive at a network that will be far more sparse, and thus far deeper, than the network $\mathcal{G}_\phi$. Of course, it is not clear whether such a network would be biologically meaningful. By undertaking a few case studies, and asking biologists to vet the graphs produced in this manner, we should be able to understand whether this approach is sensible, and/or to fine-tune the approach further.

The above 'global' question of trying to find a somewhat realistic representation of *all possible pathways* is in general not of interest to biologists. They are far more interested in 'local' questions such as the following: Suppose A and B are two genes of interest, but there is no path between them in the context-specific scaffold graph $\mathcal{G}_c$. What is the minimum number of edges from $\mathcal{G}_\phi$ that must be added so as to create a path? If there are multiple solutions to this question, which one is more realistic from a biological standpoint? There are at least two possible approaches. First, if there are two paths between genes A and B after $\mathcal{G}_c$ is augmented with edges from $\mathcal{G}_\phi$, a pathway with more 'original' edges from $\mathcal{G}_p$ and fewer 'synthetic' edges from $\mathcal{G}_\phi$ might be considered to be more realistic. Second, if the first criterion does not result in a decision, one can compute the minimum weight of all edges along a path, and call that a 'figure of merit' associated with that path. Recall from the data-processing inequality that the overall ϕ-mixing coefficient between the starting and end nodes (genes) of a path cannot be larger than the minimum of all edge weights. Thus, if there are two paths between genes A and B, we would choose the path for which the minimum edge weight is larger. However, preliminary computational experiments indicate that answering this 'local' question is not any simpler (in terms of the computational complexity) than answering the global question of making $\mathcal{G}_c$ strongly connected by adding edges from $\mathcal{G}_\phi$ in order of decreasing weight. Again, some case studies need to be undertaken to get a clearer understanding of how this approach would work in practice.

4.2 Identification of Genomic Machines

As pointed out earlier, cancer is a highly individualized disease. It is not merely that mutations in some part of the DNA cause cancer. It is also the case that mutations in other parts of the DNA have a huge impact on the responsiveness to a therapeutic regimen. Identifying which mutations cause/have caused cancer, which mutations may affect the efficacy of therapy, and tailoring the therapy appropriately, is the essence of personal medicine.

Out of the dozens of known instances, we cite just one by way of illustration [6]. The drug cetuximab is a monoclonal antibody directed against the epidermal growth factor receptor (EGFR), one of the more popular gene targets for cancer therapy. This drug is widely used as a therapy for advanced colorectal cancer, often after other forms of chemotherapy have failed. In the paper [6], the authors analyzed 394 samples of colorectal cancer to see whether they contained a mutation of the gene KRAS, which is often found to be mutated in various forms of cancer. Amongst the samples tested, 42.3 % had at least one mutation in KRAS, while the rest were 'wild type', meaning that the gene is not mutated. To paraphrase the findings of [6],

- Amongst the patients who were given best supportive care alone (i.e., no cetuximab), there was no significant difference between the survival of patients who had a KRAS mutation and those who did not.
- Amongst patients with wild-type KRAS tumors there was substantial improvement after treatment with cetuximab.
- Amongst patients with a KRAS mutation, there was no significant benefit to treatment with cetuximab.

To put it another way, a KRAS mutation does not affect survival prospects of a colorectal cancer patient who is left untreated. However, if a patient has a KRAS mutation, then cetuximab therapy is of no benefit, whereas a patient without a KRAS mutation derives significant benefit from a cetuximab treatment.

In the paper cited, the authors had a very specific hypothesis in mind, namely that KRAS mutations affected the response to cetuximab treatment. However, often the role of the computational biologists is to *generate* such hypotheses using the available data. This would entail examining the data at hand to examine not just one mutation (in this case KRAS) but multiple mutations, and assessing the significance of each possible combination of mutations. It is easy to see that if one examines k genes then there are 2^k possible states of mutations to be examined. With 400 patients (a large number in such studies), if one wishes to have an average of, say, 10 samples per state, then it is possible to examine at most $k = \lfloor \log_2(400/10) \rfloor = 5$ different genes at a time. When one undertakes very large studies involving siRNA knockdowns for example, it is not uncommon to have just a handful of samples, often in the single digits.

Therefore the emphasis in this section is on methods to identify 'genomic machines', that is, sets of genes that act in concert so as to achieve a specific function. Specifically, the situation studied is the following: First, suppose that a context-specific gene interaction network (GIN) has been constructed from expression data. Now suppose a specific perturbation is introduced, and the expression levels are measured again. The perturbation can be in the form of applying a drug, or knocking down one gene or set of genes by applying one or more siRNAs or a micro-RNA. By studying the alterations in the expression levels of individual genes, it is possible to say which genes are up-regulated (have their expression levels increased) and which genes are down-regulated by the perturbation. But this 'gene-centric' approach cannot uncover mechanisms whereby several genes, each of which show only relatively small changes in their expression levels, act together to produce a significant overall effect. Now one can ask: What does 'working together' mean? The word 'together' suggests adjacency in a biological pathway, and in the absence of detailed knowledge of such pathways, adjacency in the reverse-engineered GIN. In this section we modify an algorithm called 'Netwalk' from [7], and show how Netwalk can be combined with the ϕ-xer algorithm.

Suppose $\mathcal{G}$ is a strongly connected directed graph that comes out of the ϕ-xer algorithm. The effect of perturbations is modeled by constructing a random walk on the nodes of $\mathcal{G}$. Let n denote the number of nodes in $\mathcal{G}$ and let $\mathcal{N} := \{1, l \ldots, n\}$. To describe the random walk or Markov process $\{X_t\}$ on $\mathcal{N}$, it is necessary to specify the transition probability p_{ij} defined as

$$p_{ij} = \Pr\{X_{t+1} = j | X_t = i\}, \ \forall i, j \in \mathcal{N}.$$

The matrix P is row-stochastic in the sense that $P\mathbf{e}_n = \mathbf{e}_n$, where $\mathbf{e}$ denotes a column vector of all 1's, and the subscript denotes its dimension. To specify the matrix P, it is assumed that with each node $i \in \mathcal{N}$ there is an associated 'weight' w_i. For example, if the graph $\mathcal{G}$ has been generated from expression data of the form $\{x_{ij}, i \in \mathcal{N}, j = 1, \ldots, m\}$, then it is possible to take

$$w_i = \frac{1}{m} \sum_{j=1}^{m} x_{ij},$$

that is, the average expression level of gene i. In the absence of any other reason, it is also permissible to take all weights to be equal. Then the transition probability p_{ij} is defined as

$$p_{ij} = \begin{cases} w_j/s_i \text{ if } j \in S(i), \\ 0 \quad \text{if } j \notin S(i) \end{cases},$$

where $S(i)$ denotes the set of successor nodes of i in $\mathcal{G}$, and

$$s_i = \sum_{j \in S(i)} w_j.$$

The interpretation is that the probability of moving from node i to node j is proportional to the weight of node j, and the division by s_i serves to normalize the transitional probabilities so that they add up to one. It is easy to verify that the matrix P is row-stochastic.

The above reasoning is reminiscent of the page rank algorithm [8]. In the original page rank algorithm, the 'raw' probability defined above is modified by replacing P by the matrix

$$P \leftarrow (1 - q)P + \frac{q}{n}\mathbf{e}_n\mathbf{e}_n^T =: P^{(m)}.$$

In the case of persons browsing the Internet, the rank one correction is justified by the fact that people will often jump from the current web page to another page that is not directly connected to the current page, with some small probability q. Moreover, when they do jump, they are likely to jump to all other nodes with equal probability. This is why the rank one matrix has the special form $\mathbf{e}_n\mathbf{e}_n^T$. In the case of the Netwalk algorithm of [7], there is a similar rank one correction term, to cater to the fact that the initial graph $\mathcal{G}$ in that paper begin may not be strongly connected, and the rank one correction is supposed to model undetected interactions. However, in the present case, the graph $\mathcal{G}$ that comes out of the ϕ-xer algorithm is guaranteed to be strongly connected, so this correction term is not needed.

Once the matrix P is defined as above, one computes the stationary distribution π such that $\pi = \pi P$. Note that, since the graph $\mathcal{G}$ is strongly connected, the matrix P is irreducible, and therefore has a unique stationary distribution. One also computes the 'flow' along each edge given by

$$\mu_{ij} = \pi_i p_{ij} = \Pr\{(X_t, X_{t+1}) = (i, j)\}.$$

In the Markov chain literature, the vector μ is referred to as the doublet frequency.

The next step is to see which edges are seen to be more active as a result of the perturbation. The effect of the perturbation can be modeled by a new weight vector $\mathbf{v} = [v_i, i \in \mathcal{N}]$. The new weights can be, for example, the expression levels of all genes after the perturbation. Let $P^{(p)}$ denote the state transition matrix of the Markov chain that results from replacing the original weight vector $\mathbf{w}$ by $\mathbf{v}$, and let $\pi^{(p)}, \mu^{(p)}$ denote the associated stationary distribution and flow vectors. Then a 'figure of merit' r_{ij} is defined for each edge as

$$r_{ij} = \log \frac{\mu_{ij}^{(p)}}{\mu_{ij}}. \tag{4.1}$$

Thus $r_{ij} > 0$ if the flow along an edge is increased as a consequence of the perturbation experiment. So we can speak of individual edges as being 'up-regulated' or 'down-regulated' as a consequence of the perturbation. Finally, we can use the figure of merit to identify genomic machines, as follows: If there is a set of genes such that there is a pathway amongst them consisting of only up-regulated edges, or only down-regulated edges, then that set of genes can be said to constitute a genomic machine.

One of the advantages of the above approach is that it can be used even without any perturbations, to compare a set of data to the consensus. Suppose, as often happens, that one has a very small number of cell lines, all belonging to the same form of cancer, and that gene expression studies have been carried out all of these. Then the available data consists of a set of weights $\{w_i^l, i = 1, \ldots, n, l = 1, \ldots, k\}$, where n is the number of nodes in the graph, which is typically 20,000–30,000 genes or gene products, and k is the number of cell lines, often of the order of a dozen or so. The first step is to average the expression data among all cell lines to arrive at a 'consensus' set of weights for the overall expression study. This set of weights can be used to compute an associated set of flows μ_{ij}. Then for each index l corresponding to a particular cell line, one can construct the flows μ_{ij}^l using the associated weight vector $\{w_i^l\}$. By constructing the figure of merit r_{ij} as in (4.1), one can further carry out a longitudinal study within the cell line population.

From the above description, it is clear that the most time-consuming step in the identification of genomic machines is the computation of the stationary distribution π and the doublet frequency vector μ for several graphs, all of them having the same topology but different sets of weights for the nodes. In the original version of the page rank algorithm, the stationary distribution π is computed using the 'power method.' Since the modified matrix $P^{(m)}$ has all positive entries, the Perron theorem implies that $[P^{(m)}]^l \to \mathbf{e}_n \pi$ as $l \to \infty$. In other words, $[P^{(m)}]^l$ converges to a rank one matrix, whose rows are all equal. Consequently, for every probability vector $\mathbf{v}$, the iterated product $\mathbf{v}F^l$ converges to π (since $\mathbf{v}\mathbf{e}_n = 1$). In the present case, the matrix P corresponds to a strongly connected graph $\mathcal{G}$ and is therefore irreducible. If in addition P is also acyclic, meaning that the greatest common divisor of the lengths of all cycles in $\mathcal{G}$ is one, then it will again be the case that $P^l \to \mathbf{e}_n \pi$ as $l \to \infty$. Therefore it is once again possible to compute π using the power method.

In the case of the worldwide web, n is around eight billion and growing rapidly, so a direct implementation of the power method is not always practicable. The computer science community has developed various parallel algorithms for doing this computation. In contrast, in [9] a randomized approach is proposed for computing π. The method in [9] actually pays a lot of attention to things like ensuring synchrony of updating, communication costs etc., but we ignore these factors here. Instead we point out that, unlike in the case of the page rank algorithm and the worldwide web, the precise values of the components of π and μ are not directly relevant in biology. Rather the question of interest is whether the figure of merit r_{ij} defined in (4.1) is positive or negative for a particular edge. Once the sign of r_{ij} is determined, its magnitude is not necessarily of interest. Therefore a very germane problem in a biological context is the development of randomized algorithms for *approximate* computation of the stationary distribution and doublet frequency. The computation should be sufficiently accurate to determine the *sign* of the figure of merit for each edge (and whether its absolute value exceeds some threshold). But more is not needed, because the objects of ultimate interest are paths cycles where the edges all have the same sign, as explained earlier.

4.3 Separating Drivers from Passengers

As mentioned earlier, at present there is a massive public effort known as TCGA (The Cancer Genome Atlas) directed at extracting all relevant molecular information from every available cancerous tumor. The initial pilot studies focused on lung, brain and ovarian cancers, but recently a study of colon and rectal cancer has also been published [10]. In earlier discussions, we have focused almost exclusively on the expression levels of various genes, and used that as a way to construct context-specific networks. However, TCGA also includes much other information that could be exploited in modeling, such as copy number variation, methylation, and mutations. Mutations in specific genes lead to disruptions in the associated regulatory networks, often referred to as 'lesions'. Sequencing of tumorous tissues has thrown up and will continue to throw up a bewildering variety of mutations, some of which cause cancer (referred to as 'drivers' or 'causal mutations') while other mutations are caused by cancer (referred to as 'passengers' or 'coincidental mutations'). Various studies conducted thus far show that the frequency with which a particular gene is found to be mutated in cancerous tissue is not sufficient to distinguish the drivers of cancers from the passengers. Some additional indications need to be used to discriminate further amongst mutated genes. One possibility is to take all genes whose mutation rates are above a relatively low threshold, say 1 % of all samples tested, and by superimposing them against a context-specific genomic network, see whether there is a 'genomic machine' consisting of one or more pathways amongst a subset of these genes. Another possibility is to use the so-called developmental gene expression profile as a guide to discriminating between drivers and passengers. The objective of this section is to elaborate on this latter possibility.

We begin as usual with some background. The paper [11] presents a 'landscape' of human breast and colorectal cancer by identifying every gene that has been found in a mutated state in 11 tumor tissues of colorectal and cancer and 11 tumor tissues of breast cancer. This paper builds on an earlier work, Sjöblom et al. [12], in which 13,023 genes in 11 breast and 11 colorectal cancer tissues are analyzed. In [11], A total of 18,191 genes analyzed, out of which 1,718 were found to have at least one nonsilent mutation in either a breast or a colorectal cancer.[1] Amongst these, a total of 280 genes were identified as 'CAN-genes', that is, potentially drivers of cancer, if they had 'harbored at least one nonsynonymous mutation in both the Discovery and Validation Screens and if the total number of mutations per nucleotide sequenced exceeded a minimum threshold' [11].

It is in principle possible to carry out a very large number of experiments to test whether specific lesions are causal or not. However, in order to be definitive, it is not enough to study individual lesions—one would also have to study all possible combinations of lesions. Even if one were to focus only on the 280 CAN-genes, there would be roughly 40,000 pairs of genes, and roughly 3.6 million triplets of genes, and so on.

It is clearly impractical to carry out so many experiments. It would be preferable to have some additional indications so as to prioritize the experiments roughly in proportion to their likelihood of success. One way to achieve this is to begin with a handful of experiments where the outcomes are known, some genes being likely tumor-suppressors ('hits') while others are not likely to be so ('misses'). Then some form of pattern recognition or machine learning algorithms can be used to discriminate between the known successes ('hits') and known failures ('misses'). In the last step, this discriminating function can then be extrapoloated to all CAN-genes (or perhaps to an even larger set of genes). It must be emphasized that statistical or pattern recognition methods are not a substitute for actual experimental verification. However, by providing a high degree of separation between known hits and known misses, such methods can assist in prioritizing future experiments by increasing the likelihood of success.

Now we introduce the so-called developmental gene expression profile, and justify why it may possibly have a role in distinguishing between drivers and passengers. Development can be divided into seven stages, namely embroyd body, blastocyst,

---

[1] A nonsilent mutation is a mutation that causes a change in the amino acid sequence (primary structure) of the protein(s) produced by a gene.

fetus, neonate, infant, juvenile, and adult. The database Unigene [13] provides, for more than 100,000 genes as well as ESTs,[2] their frequency of occurrence within the tissues tested at each of the seven developmental stages. The Unigene database is far more comprehensive than earlier efforts by individual research teams to determine this type of information; see [14] for an example of this type of effort. For instance, it has been known for some time that various genes belonging to the so-called RAS family play an important role in cancer; see [15]. Out of the genes in this family, let us focus on KRAS and HRAS for now. Their Unigene entries are as follows, in parts per million:

Gene	EB	B	F	N	I	J	A
KRAS	169	80	60	0	0	54	77
HRAS	28	16	19	0	0	0	24

Since the entries are in parts per million, it is clear that these genes are not prevalent in *any* developmental stage. This raises the question as to how statistically significant the zero entries are. However, a discussion of that topic would take us too far afield.

Our hypothesis is that the developmental gene expression profile can be used to discriminate between drivers and passengers. This hypothesis is the outcome of putting together the results of a very interesting series of biological experiments. Specifically, in [16] it is shown that KRAS is essential for the development of the mouse embryo—if the KRAS gene is knocked out, then the embryo does not survive. However, as shown in [17], if the KRAS gene is not knocked out, but is instead replaced by HRAS in the KRAS locus, then the resulting HRAS-knocked in mouse embryo develops normally. Following along these lines, when HRAS was put into the KRAS locus and lung cancer was induced in these mice, the HRAS in the KRAS locus was found to be mutated, whereas the HRAS in the HRAS locus was *not mutated* [18]. Since HRAS and KRAS express themselves at different stages of the development of a mouse embryo, this observation suggests a possible relationship between the expression profile of a gene as a function of developmental stage on the one hand, and its role as a causal factor in cancer on the other hand.

To validate our hypothesis, we used another database called COSMIC (Catalogue of Somatic Mutations in Cancer) [19], that gives the observed mutation frequency of various genes in various forms of cancer. Again, COSMIC is a repository of mutation data discovered by research teams all around the world, as in [20] for example. In spite of this, since testing is expensive, not all of the roughly 30,000 known genes have been tested for mutations in all available tissues. At the moment (though of course this number keeps changing with time, albeit rather slowly), a total of 4,105 genes have been tested for mutations in any one of five forms of cancer, namely: breast, kidney, large intestine (colon), lung, and pancreas. Therefore the remaining genes were deemed not to have sufficient mutation data to permit the drawing of

---

[2] ESTs (Expressed Sequence Tags) are parts of genes that were sequenced and catalogued before whole genome sequencing became commonplace.

meaningful conclusions. Out of these 4,105 genes from COSMIC, 3,672 had entries in Unigene. These 3,672 seven-dimensional developmental gene expression profiles were clustered using the popular k-means algorithm [21]. In this approach, the given data vectors $\mathbf{x}_1, \ldots, \mathbf{x}_n \in \mathbb{R}^7$ where $n = 3672$ are clustered into k classes (k to be specified by the user) in such a way that the vectors in each class are closer to the centroid of its own class than to the centroids of all other classes. In symbols, if $\bar{\mathbf{x}}_1, \ldots, \bar{\mathbf{x}}_k$ denote the centroids of the clusters, and $\mathcal{N}_1, \ldots, \mathcal{N}_k$ denote the classes themselves, then

$$\|\mathbf{x}_i - \bar{\mathbf{x}}_k\| \leq \|\mathbf{x}_i - \bar{\mathbf{x}}_j\|, \ \forall j \neq k, \ \forall i \in \mathcal{N}_k.$$

Computing the optimal clusters is an NP-hard problem, so most often one uses some randomized algorithm. Also, the clusters themselves will be different depending on which norm is used. We have found that we get better segregation if we use the ℓ_1-norm than with the ℓ_2-norm.

Once the clusters are formed, the next step is to test whether any of these clusters is 'enriched' with known cancer drivers, compared to the remaining clusters. For determining this, it is necessary that at least a few of these 3,672 genes should be labeled, so that the problem is one of supervised learning. Fortunately, a recently completed work [22] provides a good starting point. In that paper, the authors began with the 280 CAN-genes identified by [11, 12], and identified 151 of these CAN-genes for testing in an experimental test bed that roughly approximates the environment in the colon.[3] Each of these 151 genes was individually suppressed, and the effect was observed. If the suppression of the gene resulting in cell proliferation, then the gene was labeled as a 'hit' and was presumed to play some role in colorectal cancer (CRC). If on the other hand the suppression of the gene did not result in cell proliferation, then the gene was labeled as a 'miss'. Out of the 151 genes tested, 65 turned out to be hits while the remaining 86 were labeled as misses. As a point of comparison, 400 randomly chosen genes were also tested in the same way, and only 4 were hits. Thus the fact that 65 out of 151 CAN-genes, roughly 45 %, are hits is clearly not due to chance, because out of the randomly chosen genes only 1 % were hits.

At this stage it should be pointed out that there can in fact be some ambiguity in the miss label. Even if the suppression of a particular gene did not result in cell proliferation, it is nevertheless possible that, under a different set of experimental conditions, the gene might have turned out to be a hit. From the standpoint of machine learning, this can be thought of as a problem of learning and extrapolating from labeled data, in which a positive label is 100 % accurate, whereas a negative label is treated as being inaccurate with some small probability. Learning with randomly mislabeled samples is a standard problem, and some results on this problem can be found in [23].

With the aid of these labeled genes, we then tested to see whether any of the clusters obtained by k-means was in fact enriched. Out of the 151 CAN genes tested,

---

[3] As can be imagined, this is a gross over-simplification, and the interested reader is advised to read the original paper for further details.

only 143 had entries in Unigene, so these were the labeled genes out of the 3,672 genes that were clustered. When we chose $k = 4$, the following clusters resulted.

No	Hits	Misses	Total
C1	27	47	1,807
C2	10	4	217
C3	15	16	1,016
C4	12	12	632
Total	64	79	3,672

From these results, it is apparent that Cluster No. 2 is significantly enriched for hits. The statistical significance of this was computed in two different ways. First, the null hypothesis was that the hits and misses are uniformly distributed into the four clusters, with 74, 14, 31, and 24 elements respectively, and the likelihood of there being 10 hits and 4 misses in Cluster No. 2 was computed under the assumption that the two were distributed independently. Second, the null hypothesis was that the hits are uniformly distributed into the four clusters with 1,807, 217, 2,016, and 632 elements, and the likelihood of there being 10 hits out of 217 elements in Cluster No. 2 was tested. In both tests, the null hypothesis was rejected at a 1 % level. In other words, we could assert with confidence greater than 99 % that the enrichment of hits in Cluster 2 is *not* due to chance. It is therefore plausible to conclude that several of the 217 genes in Cluster No. 2 might play a significant role in colorectal cancer.

The above clustering analysis was carried out on just 3,672 genes, because we made use of the COSMIC database for mutations in colorectal cancer. COSMIC is a compendium of results reported voluntarily by various research teams around the world. Therefore, in COSMIC, the majority of genes are not even tested for mutations in colorectal cancer. In contrast, the TCGA team has recently published a thorough analysis of colorectal cancer that includes, among other things, mutation data on a great many genes [10]. Therefore it becomes feasible to repeat the above clustering analysis on all genes that have been tested for mutations in [10], and for which developmental expression profiles exist in the Unigene database; this number would be significantly larger than 3,672. Naturally, this would result in a different set of clusters, comprising far more genes. If it were to turn out that one of those (updated) clusters is similarly enhanced by hits as reported in [22], then we could make a prediction that the genes in that cluster play a significant role in colorectal cancer. At that point, further progress would depend on finding a biologist collaborator who would be willing to invest the effort to verify the predictions. Another direction for future research would be to take the gene expression profiles reported in [10] to reverse-engineer a GIN that is specific to colorectal cancer, overlay the mutation data on top of this GIN, and see whether it is possible to identify, not just individual genes, but collections of genes (genomic machines) that play a role in colorectal cancer.

References

1. Intact: http://www.ebi.ac.uk/intact/
2. Mint: http://mint.bio.uniroma2.it/mint/welcome.do
3. Biogrid: http://thebiogrid.org/
4. String: http://thebiogrid.org/
5. Tarjan, R.E.: Depth-first search and linear graph algorithms. SIAM J. Comput. **1**(2), 146–160 (1972)
6. Karapetis, C.S., et al.: K-ras mutations and benefit from cetuximabin advanced colorectal cancer. N. Engl. J. Med. **359**(17), 1757–1765 (2008)
7. Komurov, K., White, M.A., Ram, P.T.: Use of data-biased random walkson graphs for the retrieval of context-specific networks from genomic data. PLoS Comput. Biol. **6**(8), e1000889 (2010)
8. Brin, S., Page, L.: The anatomy of a large-scale hypertextual websearch engine. Comput. Netw. ISDN Syst. **30**(10), 107–117 (1998)
9. Ishii, H., Tempo, R.: Distributed randomized algorithms for the pagerank computation. IEEE Trans. Autom. Control **55**(9), 1897–2002 (2010)
10. The Cancer Genome Atlas Network: Comprehensive molecular characterization of human colon and rectal cancer. Nature **487**, 330–337 (2012)
11. Wood, L.D., et al.: The genomic landscapes of human breast and colorectal cancers. Science **318**, 1108–1113 (2007)
12. Sjöblom, T., et al.: The consensus coding sequences of human breast and colorectal cancers. Science **314**, 268–274 (2006)
13. Unigene: http://www.ncbi.nlm.nih.gov/unigene
14. Son, C.G., et al.: Database of mrna gene expression profiles of multiple human organs. Genome Res. **15**, 443–450 (2005)
15. Bos, J.L.: ras oncogenes in human cancer: a review. Cancer Res. **49**(17), 4682–4689 (1989)
16. Koera, K., et al.: K-ras is essential for the development of the mouse embryo. Oncogene **15**(10), 1151–1159 (1997)
17. Potenza, N., et al.: Replacement of k-ras with h-ras supports normal embryonic development-despite inducing cardiovascular pathology in adult mice. EMBO Rep. **6**(5), 432–437 (2005)
18. To, M.D., et al.: Kras regulatory elements and exon 4a determine mutation specificity in lung cancer. Nat. Genet. **40**(10), 1240–1244 (2008)
19. COSMIC: http://www.sanger.ac.uk/genetics/cgp/cosmic
20. Greenman, C., et al.: Patterns of somatic mutation in human cancer genomes. Nature **132**, 153–158 (2007)
21. MacQueen, J.B.: Some methods for classification and analysis of multivariate observations. In: Proceedings of Fifth Berkeley Symposium on Mathematical Statistics and Probability, pp. 281–297. University of California Press, Berkeley (1967)
22. Eskiocak, U., et al.: Functional parsing of driver mutations in the colorectal cancer genome reveals numerous suppressors of anchorage-independent growth. Cancer Res. **71**, 4359–4365 (2011)
23. Vidyasagar, M.: Learning and Generalization: With Applications to Neural Networks and Control Systems. Springer, London (2003)

Printed by Publishers' Graphics LLC